Probability and Its Applications

Series Editors
Thomas Liggett
Charles Newman
Loren Pitt

Robert Aebi

Schrödinger Diffusion Processes

Birkhäuser Verlag
Basel · Boston · Berlin

Robert Aebi
Institut für Mathematische Statistik und Versicherungslehre
Universität Bern
Sidlerstr. 5
3012 Bern
Switzerland

A CIP catalogue record for this book is available from the Library of Congress, Washington D.C., USA

Deutsche Bibliothek Cataloging-in-Publication Data

Aebi, Robert:
Schrödinger diffusion processes /Robert Aebi. – Basel; Boston;
Berlin: Birkhäuser, 1996
        (Probability and its applications)
           ISBN-13:978-3-0348-9874-4   e-ISBN-13:978-3-0348-9027-4
           DOI: 10.1007/978-3-0348-9027-4

© 1996 Birkhäuser Verlag Basel, P.O. Box 133, CH-4010 Basel
Softcover reprint of the hardcover 1st edition 1996

Printed from the author's camera-ready manuscript on acid-free paper produced from chlorine-free pulp. TCF ∞

ISBN-13:978-3-0348-9874-4

9 8 7 6 5 4 3 2 1

Contents

5 Large Deviations

6 Interacting Diffusion Processes

7 Schrödinger Systems

Preface

'Über die Umkehrung der Naturgesetze' (On time-reversal of natural laws) –
pure curiosity made me start reading Erwin Schrödinger's nine-page-thin trea-
sure of ideas in the Reports of the Prussian Academy of Sciences for Mathematical Physics from 1931. Soon I found that not only the old-fashioned Ger-
man language but also some rather unorthodox thoughts on time-reversal and
prediction rendered immediate understanding almost impossible. Schrödinger
considers a huge cloud of independent and identical particles with known dynamics. This cloud is supposed to be observed at finite initial and final times.
Although these two observations can be quite exceptional, they are accepted as
given facts, representing reality. The problem is to determine the 'most probable' evolution of the cloud from initial to final time, conditional on the two
observations.

In this book, I intend to arrange a kind of exhibition of jewels discovered
in Schrödinger's article from 1931 and to reveal some of their consequences for
a wider field of mathematics. For the last few years I have been working on topics in probability theory, analysis, measure theory and mathematical physics
which all in a sense have their origin with Schrödinger (1931). They will be
introduced in Chapter 1, where Schrödinger's original intentions are discussed
and extended using the contemporary tools of probability theory. Quotations
from Schrödinger and Boltzmann illustrate the thinking behind these investigations, which yield an alternative view of diffusion processes. Diffusions traditionally emerge in the analysis motivated sense of Kolmogorov and Feller as
the probabilistic counterparts of semigroups. Now they are also considered in
the time-symmetrical sense of Schrödinger with their associated time-reversals.
In fact, Schrödinger's way of watching clouds means considering diffusions with
given dynamics and marginal distributions prescribed at initial and final times.
Asking for the 'most probable state' at intermediate times is actually nothing
other than predicting intermediate states from past and future data.

The mathematically crucial point in Schrödinger's investigations turns out
to be a system of two non-linear integral equations on which he comments:

> *"Die Diskussion dieses Gleichungspaares würde gewiss interes-*
> *sant, vermutlich nicht ganz leicht sein, weil es nicht linear ist. Die*
> *Existenz und Eindeutigkeit der Lösung $(\hat{\varphi}, \varphi)$ . . . halte ich wegen*
> *der Vernünftigkeit der Fragestellung, die ganz eindeutig und scharf*
> *auf diese Gleichungen führt, für ausgemacht."*

(A discussion of this pair of equations would certainly be interesting, though
presumably not entirely easy, because it is not linear. I consider the existence
and uniqueness of the solution $(\hat{\varphi}, \varphi)$ . . . to be plausible because the problem
is reasonable and yields this pair of equations directly.)

Beurling (1960) and Föllmer (1988) investigate these so-called Schrödinger systems; but Beurling's functional analysis approach assumes situations which are probabilistically too restrictive and Föllmer's large deviation approach only deals with the case of Brownian motion. Hence neither result supports Schrödinger's original hope of relating stochastic processes to Schrödinger equations. In fact, such processes must necessarily be diffusion processes with singular drift, as has been known since Nelson (1967). In Chapter 6, which has grown out of my joint work with M. Nagasawa, and in Chapter 7, Föllmer's as well as Beurling's ideas have provided the inspiration to develop methods which permit the solving of Schrödinger systems under the assumptions relevant in the theory of Schrödinger equations. Nagasawa (1989) shows that Schrödinger equations are equivalent to systems of two adjoint non-linear interacting diffusion equations. He introduces a multiplicative functional in order to establish diffusion processes related to smooth solutions of Schrödinger equations. Chapter 2, assisted by Chapters 3 and 4, provides a generalized construction of diffusion processes with singular drift determined by coefficients in a local space-time Sobolev space. This setting proves to be appropriate for the drift coefficients required by the diffusions which are related to Schrödinger equations, when treated in the standard L^2-framework.

In the course of the investigations which have led to this monograph, I have profited from a variety of stimulating circumstances. The year 1991, when I was a Swiss National Science Foundation research fellow at the Department of Mathematics at Kyoto University, provided excellent training in stochastic analysis. In 1992 as a lecturer in Edinburgh, I experienced many cold and windy days during which I had plenty of time to think about Schrödinger's ideas. Finally, it has been the generosity of the Institut für Mathematische Statistik und Versicherungslehre at the University of Berne which has allowed me to realize this book.

I would like to thank H. Föllmer, S. Ogawa, T. Shiga and H. Tanaka for their encouragement. It has been a pleasure to share M. Nagasawa's rich expertise and I am looking forward to further ambitious collaborations. The dialogue with Agnes on the essence of probability has been a constant source of inspiration.

Berne, July 1995   *Robert Aebi*

Chapter 1

Schrödinger's View of Natural Laws

This is an exploration with today's tools of probability theory into the mystery of 'Über die Umkehrung der Naturgesetze' (On time-reversal of natural laws) written by Erwin Schrödinger in 1931. He propagates the idea that diffusions given their marginal distributions at finite initial and final time are time-reversible. We are going to meet clouds of identical and independent particles considering them as realizations of diffusions. The particle dynamics is assumed to be known. Given the observation of such a cloud in terms of distribution densities at finite initial and final time, the intention is to find the 'most probable' distribution density of the cloud at intermediate times.

In the degenerate situation of Dirac-type initial and final distributions, the distribution density of the cloud at intermediate times is given by a conditional probability density which is called Schrödinger bridge. It simplifies the situation of general initial and final distribution densities in the way that only 'most probable' transitions from initial directly to final time remain to be investigated. Adopting the fundamental hypothesis of statistical mechanics which postulates that observations result from 'most probable' microscopical realizations, we find in Section 1.2 a large deviation principle for cloud transitions. A treatment by means of the Lagrangian procedure yields in Section 1.3 so-called Schrödinger systems which are pairs of non-linear integral equations. Their solutions, pairs of so-called Schrödinger multipliers, appear as weight functions in linear combinations of Schrödinger bridges which represent 'most probable' distribution densities of observed particle clouds at intermediate times. They possess a particular factorization which deserves an interpretation in terms of prediction from the past and prediction from the future.

Following Schrödinger, we will meet in Section 1.4 a striking correspondence between 'most probable' cloud densities and distribution densities $\overline{\Psi}\,\Psi$ obtained from solutions Ψ of Schrödinger equations. As a consequence, diffusions in a particular representation naturally appear as real-valued counterparts to wave functions Ψ. The so-called Schrödinger-Nagasawa representation of diffusions will be introduced in Section 1.5 as an alternative to the classical Kolmogorov-Feller representation. Traditionally, a diffusion has been given in terms of initial distribution and transition density. The Schrödinger-Nagasawa representation describes a diffusion in terms of dynamics which can be a transition density with creation and killing and a pair of Schrödinger multipliers. The advantage of this alternative representation is its time-symmetry, i.e., it contains simultaneously the representation of the time-reversed diffusion. Moreover, the Schrödinger-Nagasawa representation immediately yields the Kolmogorov-Feller representations of a diffusion and of its time-reversal. In Section 1.6 we treat the case of stochastic differential equations. The drift of a diffusion as well as the drift of its time-reversal can be identified by means of the Maruyama-Girsanov drift transformation as functions of Schrödinger multipliers.

1.1   Most probable realizations

Fixing a finite time horizon $-\infty < a < b < \infty$, we consider the partitions $(\Delta x_k)_{1 \le k \le m}$ and $(\Delta z_l)_{1 \le l \le n}$ of the state space $I\!\!R^d$ at initial time a and final time b, respectively, as well as positions $x_k \in \Delta x_k$ for $1 \le k \le m$ and $z_l \in \Delta z_l$ for $1 \le l \le n$. Hence there are m cells Δx_k at time a and n cells Δz_l at time b which describe destinations at initial and final time, respectively.

Imagine that we watch a large number N of identical and independent *particles* performing a *journey* starting at time a and ending at time b. The *observation* of the particle cloud at time a and b, respectively, can be expressed in terms of probability densities μ_a and μ_b on the state space $I\!\!R^d$. Such a watched *phenomenon* can be quite *exceptional* - we accept this as a given fact representing the *basis of our conclusions*. Schrödinger comments:

> *"Diese Beobachtung wird mehr oder weniger erstaunlich sein.*
> *Das geht uns aber hier nichts an. Wir nehmen an, dass die*
> *Verteilungen μ_a und μ_b wirklich sind und haben aus dieser Tatsache*
> *Schlüsse zu ziehen."*

(This observation will be more or less astonishing, but this does not have to bother us. We rather assume that the distributions μ_a and μ_b represent the reality on which our conclusions have to be based.)

The observation reports that $N \mu_a(x_k) \Delta x_k$ particles leave from cell Δx_k, $1 \leq k \leq m$, and that $N \mu_b(z_l) \Delta z_l$ particles arrive at cell Δz_l, $1 \leq l \leq n$. What we actually observe is the result of a *guide-book* for traveling particles containing the numbers Γ_{kl} of those particles leaving from cell Δx_k which finally arrive at cell Δz_l, $1 \leq k \leq m$, $1 \leq l \leq n$. Hence we are going to seek guide-books

$$(\Gamma_{kl})_{1 \leq k \leq m, \, 1 \leq l \leq n} \tag{1.1}$$

which obey the observation (μ_a, μ_b), i.e.,

$$N \mu_a(x_k) \Delta x_k = \sum_{l=1}^{n} \Gamma_{kl}, \quad 1 \leq k \leq m \tag{1.2}$$

$$N \mu_b(z_l) \Delta z_l = \sum_{k=1}^{m} \Gamma_{kl}, \quad 1 \leq l \leq n. \tag{1.3}$$

Dealing with *N-particle journeys*, we require *travel dynamics* of the observed kind of particles. It can be given in analytical terms by a parabolic differential operator L or in probabilistical terms by a *transition density* $p(s, x; t, y)$, $a \leq s < t \leq b$, $x, y \in \mathbb{R}^d$, where in this case, p is a (weak) fundamental solution of the *Fokker-Planck equation* $L = 0$. Thus the probability of an individual particle starting in Δx_k at time a and arriving in Δz_l at time b is given by

$$\varsigma_k \, p_{kl} \, \Delta x_k \, \Delta z_l \tag{1.4}$$

where $\varsigma_k = \varsigma(x_k)$ is any departure probability density and $p_{kl} = p(a, x_k; b, z_l)$ for $1 \leq k \leq m$, $1 \leq l \leq n$. The independence of particles enables us to compute the probability of an *N-particle journey* belonging to a certain guide-book Γ in (1.1) as

$$\prod_{k,l=1}^{m,n} \left(\varsigma_k \, p_{kl} \, \Delta x_k \, \Delta z_l\right)^{\Gamma_{kl}}.$$

In order to determine the probability of a guide-book Γ in (1.1), we have to find out how many N-particle journeys Γ contains. This means counting the arrangements of N naturally distinguishable particles as subsets of Γ_{kl} particles, $1 \leq k \leq m$, $1 \leq l \leq n$, each of which represents a certain individual journey. Their number is obtained by elementary combinatorics as

$$\binom{N}{\Gamma_{11}} \cdot \binom{N-\Gamma_{11}}{\Gamma_{12}} \cdot \binom{N-\Gamma_{11}-\Gamma_{12}}{\Gamma_{13}} \cdots$$

$$\binom{N-\sum_{l=1}^{n}\Gamma_{1l}}{\Gamma_{21}} \cdots \binom{N-\sum_{k=1}^{m}\sum_{l=1}^{n-1}\Gamma_{kl}-\sum_{k=1}^{m-1}\Gamma_{kn}}{\Gamma_{mn}}$$

$$= \frac{N!}{\prod_{k,l=1}^{m,n}\Gamma_{kl}!}. \tag{1.5}$$

Hence, provided the probabilities (1.4) of individual particle journeys, a guide-book Γ receives the probability

$$P_N(\Gamma) = N! \prod_{k,l=1}^{m,n} \frac{\left(\varsigma_k \, p_{kl} \, \Delta x_k \, \Delta z_l\right)^{\Gamma_{kl}}}{\Gamma_{kl}!} \tag{1.6}$$

We summarize that guide-books Γ in (1.1) are multinomially distributed with $m\,n$ categories having weights $\varsigma_k \, p_{kl} \, \Delta x_k \, \Delta z_l$ given in (1.4). In case of observed destination distribution densities μ_a and μ_b at initial time a and final time b, respectively, (1.2) is the continuity condition of Γ at initial time a and (1.3) is the continuity condition of Γ at final time b. Both together express that there are precisely N traveling particles and none of them goes lost on the journey described by Γ. There are many guide-books Γ which obey (1.2) and (1.3). Hence the question occurs how to determine guide-books Γ uniquely. In Sections 1.2 and 1.3 we introduce a large deviation approach and an approach in terms of systems of non-linear integral equations which are going to be investigated extensively in Chapters 6 and 7, respectively. They will finally yield the same conclusions on guide-books Γ since both are based on the *fundamental hypothesis of statistical mechanics* postulating that

*an observation on a macroscopical level is realized in
the limit of infinitely many particles by that microscopical ensemble which attains maximal probability
given the observation.*

Schrödinger (1931)'s original intention was to predict the distribution den-
sity $\rho_t(dy \mid \mu_a, \mu_b)$ of a N-particle cloud as N tends to infinity at *intermediate* times $t \in (a, b)$ for given *particle dynamics* p_{kl}, $1 \leq k \leq m$, $1 \leq l \leq n$,
in (1.4) under prescribed *marginal densities* μ_a and μ_b at initial time a and final
time b, respectively. In the present approach we first investigate the so-called
Schrödinger (Brownian) bridge (1.7) and second the *'most probable'* guide-book
density (1.16).

A Schrödinger (Brownian) bridge deals, as the notion suggests, with deterministical initial and final conditions. This degenerate situation is described
in terms of Dirac distributions as

$$\mu_a = \delta_{x_a}, \quad \mu_b = \delta_{z_b}$$

for fixed x_a, $z_b \in I\!\!R^d$. The number of particles which leave at time a from
$(x_a, x_a + dx)$ and arrive at time b in $(z_b, z_b + dz)$ is

$$N\, p(a, x_a; b, z_b)\, dx\, dz$$

and the portion

$$N\, p(a, x_a; t, y)\, dy\, p(t, y; b, z_b)\, dx\, dz$$

of them visits $(y, y + dy)$ at time t on the journey from x_a to z_b. Hence

$$\rho_t(dy \mid \delta_{x_a}, \delta_{z_b}) = \frac{p(a, x_a; t, y)\, dy\, p(t, y; b, z_b)}{p(a, x_a; b, z_b)} \tag{1.7}$$

which is a *conditional distribution density*.

In case the fundamental hypothesis of statistical mechanics has already
produced a 'most probable' guide-book in terms of a probability density q
on $I\!\!R^d \times I\!\!R^d$ for general prescribed marginal densities μ_a and μ_b, we can write
down the associated distribution density $\rho_t(dy \mid \mu_a, \mu_b)$, $t \in (a, b)$, as a q-
mixture of Schrödinger bridges (1.7) in the form

$$\rho_t(dy \mid \mu_a, \mu_b) = \int dx \int dz\, \rho_t(dy \mid \delta_x, \delta_z)\, q(x, z). \tag{1.8}$$

The two promised approaches to the 'most probable' guide-book Γ will enable us to deduce the missing term $q(x, z)$ in (1.8), and will lead to the product form of formula (1.20). This inspired Schrödinger to a *conjecture on the nature of quantum mechanics* which we will discuss in Section 1.4.

1.2   A large deviation approach

Because $(\varsigma_k\, p_{kl})_{1\leq k\leq m,\, 1\leq l\leq n}$ does not in general possess the observed, i.e., prescribed, marginal distributions $(\mu_a(x_k)\, \Delta x_k)_{1\leq k\leq m}$ and $(\mu_b(z_l)\, \Delta z_l)_{1\leq l\leq n}$, the probability (1.6) of guide-books Γ in (1.1) obeying (1.2) and (1.3) vanishes as N tends to infinity according to the law of large numbers. Hence 'most probable' under the continuity conditions (1.2) and (1.3) has to be understood as 'vanishing at slowest rate' among those satisfying (1.2) and (1.3).

Let us investigate the probability (1.6) of a guide-book Γ in (1.1) as the number N of participating particles tends to infinity. The number (1.5) of possible N-particle journeys under a guide-book Γ is going to be expressed in terms of

$$\gamma_{kl}\, \Delta x_k\, \Delta z_l \;=\; \Gamma_{kl}/N\,, \quad 1\leq k\leq m\,,\; 1\leq l\leq n\,.$$

An application of Stirling's formula

$$\nu! \;=\; (\frac{\nu}{e})^\nu\, \sqrt{2\pi\nu}\,(1+\varepsilon_\nu) \quad \text{with} \;\; \varepsilon_\nu \to 0 \;\; \text{as} \;\; \nu \nearrow \infty$$

yields

$$\frac{N!}{\prod_{k,l=1}^{m,n}\Gamma_{kl}!}$$

$$=\; \frac{(N\,e^{-1})^N\,\sqrt{2\pi N}\,(1+\varepsilon_N)}{\prod_{k,l=1}^{m,n}\{(N\,\gamma_{kl}\,\Delta x_k\,\Delta z_l\,e^{-1})^{\Gamma_{kl}}\,\sqrt{2\pi\,N\,\gamma_{kl}\,\Delta x_k\,\Delta z_l}\,(1+\varepsilon_{\Gamma_{kl}})\}}$$

$$=\; \frac{1}{(\sqrt{2\pi\,N})^{m+n-1}}\,\frac{1+\varepsilon_N}{\prod_{k,l=1}^{m,n}(1+\varepsilon_{\Gamma_{kl}})}\,\prod_{k,l=1}^{m,n}(\gamma_{kl}\,\Delta x_k\,\Delta z_l)^{-\Gamma_{kl}-\frac{1}{2}}$$

where a vanishing Γ_{kl} has factorial 1 and $\sum_{k,l=1}^{m,n}\Gamma_{kl} = N$ follows from an addition of (1.2) and (1.3). As a consequence, we can verify that the guide-book probability (1.6) possesses the limit

$$\lim_{N\nearrow\infty}\frac{1}{N}\,\log P_N(\Gamma) = \tag{1.9}$$

$$= \lim_{N \nearrow \infty} \frac{1}{N} \left(\sum_{k,l=1}^{m,n} \{ \log(\varsigma_k \, p_{kl}) - \log \gamma_{kl} \} \, N \, \gamma_{kl} \, \Delta x_k \, \Delta z_l - \frac{1}{2} \sum_{k,l=1}^{m,n} \log \frac{\Gamma_{kl}}{N} \right.$$

$$\left. - (m+n-1) \log \sqrt{2\pi N} + \log(1 + \varepsilon_N) - \sum_{k,l=1}^{m,n} \log(1 + \varepsilon_{\Gamma_{kl}}) \right)$$

$$= -H_\Delta(\gamma_{..} \mid \varsigma. \, p_{..})$$

where

$$H_\Delta(\gamma_{..} \mid \varsigma. \, p_{..}) = \sum_{k,l=1}^{m,n} \log\left(\frac{\gamma_{kl}}{\varsigma_k \, p_{kl}}\right) \gamma_{kl} \, \Delta x_k \Delta z_l \qquad (1.10)$$

is called the *relative entropy* of γ with respect to $\varsigma \, p$ in discrete situations. In case γ is not absolutely continuous with respect to $\varsigma \, p$, i.e.,

$$\varsigma_k \, p_{kl} = 0 \Rightarrow \gamma_{kl} = 0 \quad \text{for} \ \ 1 \le k \le m, \quad 1 \le l \le n$$

is violated, $H_\Delta(\gamma_{..} \mid \varsigma. \, p_{..})$ is set to be infinity. We will meet relative entropies in Chapters 5 and 6. In statistical mechanics they are considered as the natural non-negative quantities to describe the amount of randomness in particle systems and were already discussed by Boltzmann (1896) in his lectures on the theory of gas.

The limit (1.9) shows that the probability (1.6) vanishes *exponentially fast* as the number of participating particles N tends to infinity. Hence (1.9) represents a so-called large deviation principle with the rate function (1.10). Seeking the 'most probable' Γ in (1.1) under the continuity conditions (1.2) and (1.3) means to minimize (1.10) under

$$\sum_{l=1}^{n} \gamma_{kl} \, \Delta z_l = \mu_a(x_k), \qquad 1 \le k \le m \qquad (1.11)$$

$$\sum_{k=1}^{m} \gamma_{kl} \, \Delta x_k = \mu_b(z_l), \qquad 1 \le l \le n. \qquad (1.12)$$

In the continuous situation of infinitely many cells of infinitesimal size, the combinatorics employed to get (1.5) fail to work. Hence 'most probable' is proposed to be defined in analogy to the discrete situation of a minimal (1.10) under (1.11) and (1.12). We call a guide-book density q on $\mathbb{R}^d \times \mathbb{R}^d$ 'most probable' under the observation (μ_a, μ_b), if it is a solution of the variational principle

$$\min_{\gamma:\left\{\begin{array}{rcl} \int dz\,\gamma(x,z) &=& \mu_a(x) \\ \int dx\,\gamma(x,z) &=& \mu_b(z) \end{array}\right.} H(\gamma \mid \varsigma\,p) \qquad (1.13)$$

where

$$H(\gamma \mid \varsigma\,p) = \int dx \int dz\,\gamma(x,z)\,\log\left(\frac{\gamma(x,z)}{\varsigma(x)\,p(x,z)}\right) \qquad (1.14)$$

which is the continuous analog of (1.10). Csiszar (1975, 1984) develops a kind of geometry in the space of probability measures employing the relative entropy $H(\,.\mid.\,)$ as a 'distance'. The minimal point q of (1.13) appears as the so-called I-projection of $\varsigma\,p$ onto the convex and variation closed subset determined by the continuous analog of (1.11) and (1.12). However, the relative entropy is not symmetrical in its arguments, hence a priori it cannot be a metric. As the rate function of the large deviation principle (1.9) associated with (1.13), it approximately describes the decreasing probability of an observation (μ_a,μ_b) by

$$e^{-N\,H(q\mid\varsigma\,p)}$$

as the number N of traveling particles increases to infinity. Chapter 6 deals with a large deviation principle associated with (1.13) for dynamics (1.4) with singular creation and killing. Such dynamics is motivated from quantum mechanics as we will learn in Section 1.4 and Chapter 2. The investigations of a Sanov-type large deviation principle in Section 6.2 will lead to the phenomenon of propagation of chaos in entropy. It can be employed to understand quantum mechanics in terms of statistical mechanics.

1.3 Prediction from past and future

Minimizing $H_\Delta(\gamma_{..} \mid \varsigma.\,p_{..})$ in (1.10) under the continuity conditions (1.11) and (1.12) permits the Lagrangian procedure. We consider

$$L_\Delta(\gamma_{..},\lambda_{..}) = \sum_{k,l=1}^{m,n} \log\left(\frac{\gamma_{kl}}{\varsigma_k\,p_{kl}}\right)\gamma_{kl}\,\Delta x_k\,\Delta z_l \qquad (1.15)$$

$$- \sum_{k=1}^{m}\lambda_{k.}\left(\sum_{l=1}^{n}\gamma_{kl}\,\Delta z_l - \mu_a(x_k)\right)\Delta x_k$$

$$- \sum_{l=1}^{n}\lambda_{.l}\left(\sum_{k=1}^{m}\gamma_{kl}\,\Delta x_k - \mu_b(z_l)\right)\Delta z_l$$

for $m+n$ Lagrangian multipliers $\lambda_{k\cdot}$ and $\lambda_{\cdot l}$. The first $m\,n$ Lagrangian equations

$$\frac{\partial L_\Delta}{\partial \gamma_{ij}} \;=\; (1 + \log(\frac{\gamma_{ij}}{\varsigma_i\, p_{ij}})) - \lambda_{i\cdot} - \lambda_{\cdot j})\,\Delta x_i\,\Delta z_j = 0$$

yield

$$e\,\frac{\gamma_{ij}}{\varsigma_i\, p_{ij}} \;=\; \exp\{\lambda_{i\cdot}\}\,\exp\{\lambda_{\cdot j}\}$$

for $1 \le i \le m$, $1 \le j \le n$. Hence, the 'most probable' guide-book probability density takes the form

$$q_{ij} = \hat{\varphi}_i\, \varsigma_i\, p_{ij}\, \varphi_j \tag{1.16}$$

on $(\Delta x_k)_{1\le k\le m} \times (\Delta z_l)_{1\le l\le n}$, where

$$\hat{\varphi}_i = \exp\{\lambda_{i\cdot}\}, \quad 1 \le i \le m$$

depends exclusively on *the past at time a*, and

$$\varphi_j = \exp\{\lambda_{\cdot j} - 1\}, \quad 1 \le j \le n$$

depends exclusively on *the future at time b*. We call the non-negative $(\hat{\varphi}_i)_{1\le i\le m}$ and $(\varphi_j)_{1\le j\le n}$ Schrödinger multipliers and refer to Section 2.2 for a general discussion. They are up to multiplicative constants unique solutions of the so-called *Schrödinger system*

$$\hat{\varphi}_k\, \varsigma_k \sum_{l=1}^{n} p_{kl}\, \varphi_l\, \Delta z_l \;=\; \mu_a(x_k), \quad 1 \le k \le m \tag{1.17}$$

$$\sum_{k=1}^{m} \hat{\varphi}_k\, \varsigma_k\, p_{kl}\, \Delta x_k\, \varphi_l \;=\; \mu_b(z_l), \quad 1 \le l \le n \tag{1.18}$$

which represents the Lagrangian equations $\partial L_\Delta / \partial \lambda_{i\cdot} = 0$ and $\partial L_\Delta / \partial \lambda_{\cdot j} = 0$ expressed in terms of (1.16). (1.17) & (1.18) form a discrete version of a system of non-linear integral equations deduced from the continuity conditions (1.11) and (1.12).

The investigation of diffusion processes requires guide-book densities q in (1.16) in a continuous setting. In the limit of infinitesimal cell size however, any refinement of the discretization $(\Delta x_k)_{1\le k\le m}$, $(\Delta z_l)_{1\le l\le n}$ yields an infinite number of equations corresponding to (1.11) and (1.12) and the Lagrangian procedure can no longer be applied. Föllmer (1988) and Aebi-Nagasawa (1992a) prove for Brownian motion dynamics p in (1.4) and for dynamics p with singular creation and killing, respectively, that solutions of the large deviation principle (1.13) are solutions of the continuous version of the Schrödinger system (1.17) & (1.18) and vice versa. As a consequence of Csiszar (1975, 1984), guide-book densities q possess the characteristical factorization

$$q(x, z) = \hat{\varphi}(x) \, \varsigma(x) \, p(a, x; b, z) \, \varphi(z) \quad \text{for} \ \ x, z \in I\!\!R^d \qquad (1.19)$$

in terms of Schrödinger multipliers $\hat{\varphi}$ and φ. Chapter 6 will provide these relations in a general framework. Existence and uniqueness of solutions of Schrödinger systems (1.17) & (1.18) were not discussed by Schrödinger himself. Such questions were investigated by Bernstein (1932), Fortet (1940), Beurling (1960), Jamison (1975) and Aebi (1995b); the last one deals with more general dynamics p which will be treated in Chapter 7.

As a consequence of (1.8) in terms of (1.7) and (1.19), the distribution density ρ_t of an observed cloud of infinitely many independent and identical particles at intermediate times $t \in (a, b)$ is given by

$$\rho_t(dy \mid \mu_a, \mu_b) \qquad (1.20)$$
$$= \int dx \int dz \, \frac{p(a, x; t, y) \, dy \, p(t, y; b, z)}{p(a, x; b, z)} \, \hat{\varphi}(x) \, \varsigma(x) \, p(a, x; b, z) \, \varphi(z)$$
$$= \hat{\varphi}(t, y) \, \varphi(t, y) \, dy$$

where

$$\varphi(t, y) \ = \ \int p(t, y; b, z) \, dz \, \varphi(z) \,, \quad t \in [a, b) \qquad (1.21)$$

$$\hat{\varphi}(t, y) \ = \ \int \hat{\varphi}(x) \, \varsigma(x) \, dx \, p(a, x; t, y) \,, \quad t \in (a, b] \qquad (1.22)$$

for $y \in I\!\!R^d$, are *space-time harmonic* and *space-time co-harmonic* functions of $p(s, x; t, y)$, $a \leq s < t \leq b$, respectively.

The *structures of the distribution density* ρ_t in (1.20) clarify the nature of 'most probable' diffusions for given dynamics p under prescribed marginal densities μ_a and μ_b at initial time a and final time b, respectively. In fact, let us consider the Schrödinger multipliers $\hat{\varphi}$ and φ as the *'history' from past and future*, respectively, determined by the observation (μ_a, μ_b) through the Schrödinger system (1.17)&(1.18). According to (1.22), $\hat{\varphi}(t, .)$ is the *prediction from the past* and according to (1.21), $\varphi(t, .)$ is the *prediction from the future*, i.e., time-reversed prediction. Hence the distribution density ρ_t at intermediate times $t \in (a, b)$ is the *product of prediction from the past and prediction from the future*. We should heed Schrödinger (1931)'s general conclusions:

"Die sogenannten irreversiblen Naturgesetze zeichnen, wenn man sie statistisch deutet, eigentlich keine Zeitrichtung aus. Denn was sie im speziellen Fall aussagen, hängt nur von den zeitlichen Grenzbedingungen in zwei 'Querschnitten' a und b ab und ist bezüglich dieser Querschnitte vollkommen symmetrisch, ohne dass es auf deren zeitliche Reihenfolge irgendwie ankäme. Das wird nur dadurch etwas verschleiert, dass wir im allgemeinen bloss einen der beiden Querschnitte als wirklich beobachtet ansehen, während für den anderen die zuverlässige Regel gilt, dass, wenn er in hinreichenden zeitlichen Abstand verlegt wird, der Zustand grösster Unordnung oder maximaler Entropie dort angenommen werden darf. Dass diese Regel das Richtige trifft, ist eigentlich sehr merkwürdig und, wie ich glaube, nicht logisch deduzierbar. Aber jedenfalls zeichnet auch sie keinen Zeitsinn aus, denn sie gilt gleichmässig, in welche von den beiden Zeitrichtungen man auch den zweiten Querschnitt verlegt, wenn er bloss zeitlich genügend weit von dem ersten entfernt ist."

(If considered from a statistical point of view, the so-called irreversible natural laws do not actually determine a specific direction of time. In fact, their statements only depend on the space cross-sections at two points of time a and b, and they are entirely symmetrical with respect to the conditions at time a and b. This is somewhat obscured by our habit of observing only one cross-section in space. The second we usually put off to a sufficiently remote time, so we are entitled to assume that the state of biggest randomness, i.e., of maximal entropy, is achieved. It is really very strange that this rule should be true. I believe that it cannot be deduced logically. Anyway, this rule does not determine a specific direction of time either. It holds equally for both directions of time provided only that the second space cross-section is sufficiently distant in time from the first one.)

In Schrödinger's conception of diffusions, their *time-reversals* are always considered simultaneously. As witness he cites Boltzmann who gives in 1898 the philosophical comment:

Dass eine Welt ebensogut denkbar wäre, in welcher alle Naturvorgänge in verkehrter Reihenfolge ablaufen würden, unterliegt keinem Zweifel; jedoch hätte ein Mensch, welcher in dieser verkehrten

Welt leben würde, keineswegs ein andere Empfindung als wir. Er
würde eben das, was wir Zukunft nennen, als Vergangenheit und
'umgekehrt' bezeichnen.

(There is no doubt that a world could also be imagined in which all natural processes happen in time-reversed order. A human being living in such a reversed world would not feel differently from us. He would simply call past what we call future and vice versa.)

1.4   An analogy to wave functions

Schrödinger stresses in 1931 that for him, the most interesting aspect of the distribution density ρ_t, $t \in [a, b]$, as a product (1.20) is its formal analogy to distribution densities $\overline{\Psi}_t \Psi_t$ formed out of solutions Ψ_t of Schrödinger equations and their complex conjugates $\overline{\Psi}_t$, $t \in [a, b]$. Let us follow his discussion ending with a quotation of his conclusions.

Schrödinger remarks that ρ_t in (1.20) cannot be a solution of the parabolic differential equation

$$L = 0 \tag{1.23}$$

because in this case, ρ_t would be predicted for $t \in [a, b)$ by μ_b exclusively, i.e.,

$$\rho_t^{\mathrm{pred}\ L}(y) = \int p(t, y; b, z)\, dz\, \mu_b(z) \quad \text{for } y \in \mathbb{R}^d.$$

Similarly, ρ_t cannot be a solution of

$$L^* = 0 \tag{1.24}$$

where L^* is the adjoint operator of L, because in this case, ρ_t would be predicted for $t \in (a, b]$ by μ_a exclusively, i.e.,

$$\rho_t^{\mathrm{pred}\ L^*}(y) = \int \mu_a(x)\, dx\, p(a, x; t, y) \quad \text{for } y \in \mathbb{R}^d.$$

Is there something wrong with the problem of finding ρ_t? First we notice two situations with immediate solutions. One is the Schrödinger bridge (1.7) which however does not provide any $\hat{\varphi}$, φ. In the other situation, the observation (μ_a, μ_b) is supposed to be just $\mu_a(x) = \varsigma(x)$ for $x \in \mathbb{R}^d$ and

$$\mu_b(z) = \int \mu_a(x)\, dx\, p(a, x; b, z) \quad \text{for } z \in \mathbb{R}^d$$

which corresponds to the ordinary diffusion phenomenon governed by (1.24). Then the Schrödinger multipliers turn out to be

$$\hat{\varphi} = k, \qquad \varphi = 1/k \qquad \text{for constants } k > 0.$$

Second, the structures of ρ_t, $t \in [a, b]$, are in a way quite obvious, however unusual in diffusion theory so far. (1.20) is the product of any solution of (1.23) and any solution of (1.24). Actually, $\varphi(t, y)$ in (1.21) is nothing else than the solution of (1.23) determined by $\varphi(b, z)$ and $\hat{\varphi}(t, y)$ in (1.22) is nothing else than the solution of (1.24) determined by $\hat{\varphi}(a, x)$, both represented in terms of the fundamental solution $p(s, x; t, y)$ of the corresponding Fokker-Planck equation (1.23) for $s, t \in [a, b]$, $s < t$, $x, y \in \mathbb{R}^d$. If a product of solutions of (1.23) and (1.24), say $\hat{\varphi}(t, y)\, \varphi(t, y)$, has a finite space integral for at least one t-value, then it can be employed due to normalization as a distribution density at time $t \in [a, b]$. In fact, (1.21) and (1.22) yield

$$\frac{\partial}{\partial t} \int dy\, \hat{\varphi}(t, y)\, \varphi(t, y)$$

$$= \frac{\partial}{\partial t} \int dy \int \hat{\varphi}(x)\, \varsigma(x)\, dx\, p(a, x; t, y) \int p(t, y; b, z)\, dz\, \varphi(z)$$

$$= \frac{\partial}{\partial t} \int \int \hat{\varphi}(x)\, \varsigma(x)\, dx\, p(a, x; b, z)\, dz\, \varphi(z) = 0.$$

As a familiar example we consider the free (Brownian) particle. Its law of motion is provided by the space-time generator

$$L = \frac{\partial}{\partial t} + \frac{1}{2}\, \triangle. \tag{1.25}$$

The distribution density of the free particle in terms of ρ_t, $t \in [a, b]$, in (1.20) is a product of a solution (1.21) of the Fokker-Planck equation $L = 0$ and a solution (1.22) of the Fokker-Planck equation $L^* = 0$. As a consequence, ρ_t, $t \in [a, b]$, does not evolve in a specified direction of time. In quantum mechanics, the distribution density of the free particle is given as $\overline{\Psi}_t \Psi_t$, $t \in [a, b]$, which is a product of a solution Ψ_t of the Schrödinger equation

$$\sqrt{-1}\, \frac{\partial \Psi}{\partial t} + \frac{1}{2}\, \triangle \Psi = 0$$

and a solution $\overline{\Psi}_t$ of the adjoint Schrödinger equation

$$-\sqrt{-1}\,\frac{\partial \Psi}{\partial t} + \frac{1}{2}\,\triangle\,\Psi = 0.$$

Both ρ_t and $\overline{\Psi}_t\,\Psi_t$ are obtained as products in a bilinear way from linear differential equations of first order in time and second order in space. Both describe a conservative and time-reversible phenomenon. This in contrast to the traditional distribution density satisfying the Fokker-Planck equation $L^* = 0$ which describes a dissipative and time-irreversible phenomenon. Schrödinger concludes his investigations by confessing:

> *"Ob die Analogie sich zur Klärung quantenmechanischer Begriffe nützlich erweisen wird, vermag ich noch nicht vorauszusehen. Die oben erwähnte $\sqrt{-1}$ bedeutet selbstverständlich trotz alledem einen sehr tiefgreifenden Unterschied. Ich kann mich nicht enthalten, hier einigen Worten A.S. Eddingtons über die Interpretation der Wellenmechanik Raum zu geben - so dunkel sie auch sind - sie stehen S. 216f. seiner Gifford-lectures ('The nature of the physical world', Cambridge 1928):"*

> > *"Die ganze Interpretation ist sehr dunkel, scheint aber davon abzuhängen, ob es sich handelt um die Wahrscheinlichkeit zum Zwecke einer Voraussage. $\overline{\Psi}\,\Psi$ wird erhalten, indem man zwei symmetrische Systeme von Ψ-Wellen einführt, die in entgegengesetzter Zeitrichtung wandern; das eine von ihnen hat vermutlich etwas zu tun mit einem Wahrscheinlichkeitsschluss aus dem bekannten (oder als bekannt vorausgesetzten) Zustand des Systems in einem späteren Zeitpunkt."*

(I am not able to predict whether this analogy will provide a better understanding of notions in quantum mechanics. The mentioned $\sqrt{-1}$ means of course a significant difference. I cannot resist to quote A.S. Eddington giving his rather vague interpretation of wave functions on p. 216f. in his Gifford-lectures.) Here we can refer to Arthur Eddington's original statement:

> *"The whole interpretation is very obscure, but it seems to depend on whether you are considering the probability after you know what*

has happened or the probability for the purpose of prediction. The
$\overline{\Psi}\,\Psi$ *is obtained by introducing two symmetrical systems of Ψ waves*
traveling in opposite directions in time; one of these must presum-
ably correspond to probable inference from what is known (or stated)
to have been the condition at a later time."

In Chapter 2 we show that a Schrödinger equation with scalar as well
as vector potential is equivalent to a pair of adjoint non-linear interacting diffusion equations which determine mutually time-reversed diffusion processes
with in general singular drift. Since the theory of Schrödinger equations has
been developed in a L^2-framework, differentiability of wave functions Ψ must
not be relied on when seeking diffusion processes related to Schrödinger equations. Hence we will employ the local Sobolev space for Schrödinger multipliers
defined in Aebi (1993).

1.5   Two representations of diffusions

This section features a representation of diffusion processes which started to occur in Schrödinger's view of diffusions treated in Section 1.1 and which is essential in Nagasawa (1989)'s investigations of diffusions related to wave functions.
For bounded, non-negative, measurable functions $\varphi(t,y) \in C^{1,2}([a,b] \times I\!R^d)$
(or in the local space-time Sobolev space $H_{loc}^{1,2}$ and jointly continuous, see Section 2.2) we define a so-called *reference potential*

$$c(t,y) = -\frac{L\varphi(t,y)}{\varphi(t,y)} \tag{1.26}$$

on the subset

$$D = \{(t,y) \in [a,b] \times I\!R^d : \varphi(t,y) > 0\} \tag{1.27}$$

where L is any parabolic differential operator as considered in Section 2.1. The
diffusion equation

$$(L+c)\,p = 0 \tag{1.28}$$

with in general singular creation and killing potential c in (1.26) possesses
a (weak) fundamental solution $p(s,x;t,y) \in (H_{loc}^{1,2})^2$ which will be deduced
in Section 2.3. Since the non-negative p satisfies the Chapman-Kolmogorov
equation, we can employ it as a general form of dynamics. However, we have to

notice that p can vanish on sets with positive Lebesgue measure and that it is in general not normalized, i.e., not a probability transition density. The function φ trivially satisfies (1.28) because of (1.26). Hence it can be represented in terms of the fundamental solution p of (1.28) as

$$\varphi(s,x) = \int p(s,x;t,y)\, dy\, \varphi(t,y), \quad a < s < t < b,\ x \in \mathbb{R}^d \tag{1.29}$$

which means that φ is space-time harmonic with respect to p. It is easily verified that q defined by means of 'conjugation' of p with φ, i.e.,

$$q(s,x;t,y) = \frac{1}{\varphi(s,x)}\, p(s,x;t,y)\, \varphi(t,y) \tag{1.30}$$

for $s < t$, $(s,x),(t,y) \in D$ in (1.27), is a probability transition density.

The function $\varphi(t,y)$, $(t,y) \in [a,b] \times \mathbb{R}^d$, is joined by a bounded, non-negative, measurable function $\hat{\varphi}(a,\,.\,)$ on $\mathbb{R}^d$ in order to specify an initial probability distribution density of the form $\hat{\varphi}(a,\,.\,)\,\varphi(a,\,.\,)$. We extend $\hat{\varphi}$ as a space-time co-harmonic function of the fundamental solution p of (1.28), i.e.,

$$\hat{\varphi}(t,y) = \int \hat{\varphi}(a,x)\, dx\, p(a,x;t,y) \tag{1.31}$$

for $(t,y) \in (a,b] \times \mathbb{R}^d$. Hence $\hat{\varphi}$ is a (weak) solution of

$$(L^* + c)\, p = 0$$

which is the adjoint equation of (1.28).

Probabilistically, we employ p, $\hat{\varphi}$ and φ to define a time-inhomogeneous *conservative*, i.e., mass preserving, diffusion process

$$(X_t,\ a \le t \le b,\ Q) \tag{1.32}$$

by its finite dimensional distributions as

$$\begin{aligned}
&Q[f(X_a, X_{t_1}, \ldots, X_{t_n}, X_b)] \\
&= \int \hat{\varphi}(a,x)\, \varphi(a,x)\, dx\, q(a,x;t_1,y_1)\, dy_1\, q(t_1,y_1;t_2,y_2)\, dy_2 \cdots . \tag{1.33} \\
&\qquad \cdot q(t_n,y_n;b,z)\, dz\, f(x,y_1,\ldots,y_n,z) \\
&= \int \hat{\varphi}(a,x)\, dx\, p(a,x;t_1,y_1)\, dy_1\, p(t_1,y_1;t_2,y_2)\, dy_2 \cdots . \tag{1.34} \\
&\qquad \cdot p(t_n,y_n;b,z)\, dz\, \varphi(b,z)\, f(x,y_1,\ldots,y_n,z)
\end{aligned}$$

for bounded and measurable functions $f(x, y_1, \ldots, y_n, z)$ on $(I\!\!R^d)^{n+2}$ where $a < t_1 < \ldots < t_n < b$. Traditionally, (1.33) is called the *Kolmogorov-Feller representation* of the diffusion process (1.32) in terms of initial distribution density $\hat{\varphi}(a, \cdot)\, \varphi(a, \cdot)$ and probability transition density (1.30). As an alternative, we introduce in the spirit of Schrödinger (1931) and Nagasawa (1989) the *Schrödinger-Nagasawa representation* (1.34) of the diffusion process (1.32) in terms of dynamics p in (1.4) with creation and killing and the Schrödinger multipliers $\hat{\varphi}$ and φ.

The alternative representation (1.34) yields

$$
\begin{aligned}
&Q[f(X_a, X_{t_1}, \ldots, X_{t_n}, X_b)] \\
=\ &\int dx\, \hat{\varphi}(a, x)\, p(a, x; t_1, y_1)\, \frac{1}{\hat{\varphi}(t_1, y_1)}\, dy_1\ \cdot \\
&\cdot\, \hat{\varphi}(t_1, y_1)\, p(t_1, y_1; t_2, y_2)\, \frac{1}{\hat{\varphi}(t_2, y_2)}\, dy_2 \cdot \ldots \\
&\cdot\, p(t_n, y_n; b, z)\, \frac{1}{\hat{\varphi}(b, z)}\, \hat{\varphi}(b, z)\, \varphi(b, z)\, dz\, f(x, y_1, \ldots, y_n, z)
\end{aligned}
\tag{1.35}
$$

in terms of $\hat{\varphi}$ given in (1.31). By defining

$$
\hat{q}(s, x; t, y) = \hat{\varphi}(s, x)\, p(s, x; t, y)\, \frac{1}{\hat{\varphi}(t, y)} \quad \text{for } s < t
\tag{1.36}
$$

on the subset

$$
\hat{D} = \{(t, y) \in [a, b] \times I\!\!R^d : \hat{\varphi}(t, y) > 0\}
\tag{1.37}
$$

the representation (1.35), read from the right to the left with t running backwards from b to a, describes (1.32) in terms of its time-reversed probability transition density (1.36) and its final distribution density $\hat{\varphi}(b, \cdot)\, \varphi(b, \cdot)$. Thus (1.35) is the Kolmogorov-Feller representation of the time-reversed diffusion process (1.32). We notice how nicely the Schrödinger-Nagasawa representation (1.34) links the Kolmogorov-Feller representations (1.33) and (1.35) of a diffusion and its time-reversal. This illustrates well the time-symmetry of a Schrödinger-Nagasawa representation where its construction can also be started with a bounded, non-negative $\hat{\varphi} \in C^{1,2}([a, b] \times I\!\!R^d)$ (or in $H_{loc}^{1,2}$ and jointly continuous) and a bounded, non-negative, measurable $\varphi(b, \cdot)$ on $I\!\!R^d$ such that $\hat{\varphi}(b, \cdot)\, \varphi(b, \cdot)$ is a probability distribution density.

The state space of a diffusion process (1.32) in Schrödinger-Nagasawa representation can be described by $D \cap \hat{D}$ defined in (1.27) and (1.37), respectively.

In fact, the distribution density of (1.32) at time $t \in (a, b)$ takes the form

$$
\begin{aligned}
\rho_t(dy) &= \int \hat{\varphi}(a, x)\, \varphi(a, x)\, dx\, q(a, x; t, y)\, dy \\
&= \int \hat{\varphi}(a, x)\, dx\, p(a, x; t, y)\, dy \int p(t, y; b, z)\, dz\, \varphi(b, z) \\
&= \hat{\varphi}(t, y)\, \varphi(t, y)\, dy
\end{aligned}
\tag{1.38}
$$

which is precisely (1.20) where $\varphi(t, .)$ in (1.29) and $\hat{\varphi}(t, .)$ in (1.31) correspond to (1.21) and (1.22), respectively. As we revealed in Section 1.3, the factorized form of (1.38) allows an interpretation in terms of *prediction from past and future*. Let us agree on the terminology that diffusion processes in Schrödinger-Nagasawa representation are called Schrödinger processes from now on. Nevertheless we are always aware that they are general diffusion processes represented in a specific time-symmetrical way.

Remark 1.5.1 As was motivated in Section 1.4 and as will be deduced in Chapter 2, so-called Schrödinger processes should be considered as the real-valued counterparts to solutions of Schrödinger equations . Nagasawa (1993) develops an entire quantum physics approach built up on Schrödinger's and Kolmogorov's ideas of diffusion processes. He discusses probabilistical view-points of quantum mechanics due to Fényes, Nelson, Bohm and others on a common basis and finally, he establishes Schrödinger equations as a part of diffusion theory.

1.6   Identification of drift

We are going to reveal the roles of dynamics p and the Schrödinger multipliers $\hat{\varphi}$ and φ in stochastic differential equations. In a first step the example of Brownian motion dynamics p is considered. It causes comparably simple computations in terms of the well known *Maruyama-Girsanov drift transformation* where we will assume that the Schrödinger multipliers are in $C^{1,2}([a, b] \times I\!\!R^d)$. A further step provided by Section 2.4 will consist of the treatment of dynamics p which may vanish on sets with positive Lebesgue measure and which can possess singular creation and killing. Dealing with stochastic differential equations, diffusion processes are considered as measures on the space of continuous paths. For any $X \in C([a, b], I\!\!R^d)$ let us denote by X_t the state of the path X at time $t \in [a, b]$. Accordingly, our dynamics p can be represented by Wiener measures $W^{(r)}$, $r = a, b$, with initial distribution density q_r at time a. In case of a probability measure R on $C([a, b], \mathbb{R}^d)$, the continuity conditions corresponding to (1.2) and (1.3) are given by

$$R(X_a \in dx) \;\; = \;\; \mu_a(dx) \tag{1.39}$$
$$R(X_b \in dz) \;\; = \;\; \mu_b(dz). \tag{1.40}$$

For regular versions of conditional probabilities of R conditioned on initial and final states we write

$$R_x^z = R(\,.\, \mid X_a = x, X_b = z).$$

The associated 2-dimensional marginal distributions at initial and final time are denoted by

$$\nu^{(r)}(dx\,dz) \;\; = \;\; W^{(r)}((X_a, X_b) \in dx\,dz)\,, \quad r = a, b$$
$$\gamma(dx\,dz) \;\; = \;\; R((X_b, X_b) \in dx\,dz)$$

where γ is assumed to possess the density $\gamma(x, z)$ and $\nu^{(r)}$ can be represented as

$$\nu^{(r)}(dx\,dz) = q_r(x)\,p(a, x; b, z)\,dx\,dz\,, \quad r = a, b \tag{1.41}$$

in terms of the Gaussian kernel $p(a, x; b, z)$.

Let us formulate a large deviation principle for probability measures on the path space $C([a, b], \mathbb{R}^d)$ along the lines of Section 1.3. Because of

$$\frac{dR}{dW^{(a)}}(X) = \frac{d\gamma}{d\nu^{(a)}}(X_a, X_b)\,\frac{dR_{X_a}^{X_b}}{dW_{X_a}^{X_b}}(X)\,, \quad W^{(a)}\text{-a.s.}$$

the relative entropy expression extending (1.14) to measures on $C([a, b], \mathbb{R}^d)$ takes the form

$$H(R \mid W^{(a)}) \;\; = \;\; \int \log\!\Big(\frac{dR}{dW^{(a)}}(X)\Big)\,dR(X) \tag{1.42}$$
$$= \;\; \int\!\!\int \log\!\Big(\frac{d\gamma}{d\nu^{(a)}}(x, z)\Big)\,\gamma(dx\,dz)$$
$$+ \;\; \int\!\!\int\!\!\int \log\!\Big(\frac{dR_x^z}{dW_x^z}(X)\Big)\,dR_x^z(X)\,\gamma(dx\,dz).$$

For probability measures R_i, $i = 1, 2$, the relative entropy $H(R_1 \mid R_2)$ is always non-negative and vanishes if and only if $R_1 = R_2$, R_1-almost certainly.

Consequently, a minimum of (1.42) can only be attained if

$$R_x^z = W_x^z, \quad W_x^z\text{-a.s. for } \gamma\text{-almost all } (x, z) \in \mathbb{R}^{2d}.$$

Hence the variational problem

$$\min_{R \text{ with } (1.39) \text{ and } (1.40)} H(R \mid W^{(a)}) \tag{1.43}$$

is equivalent to (1.13) and the minimal point is a Schrödinger process which can be given in terms of a probability measure Q on $C([a, b], \mathbb{R}^d)$ as

$$dQ(X) = q(X_a, X_b) \, dW_x^z(X) \, dx \, dz. \tag{1.44}$$

The goal of this section is to determine the drift vector of a Schrödinger process $(X_t, \ a \le t \le b, \ Q)$ and the drift vector of its time-reversed process $(\hat{X}_t = X_{a+b-t}, \ a \le t \le b, \ Q)$. The Radon-Nikodym derivatives of the Schrödinger process Q with respect to $W^{(r)}$, $r = a, b$, are computed based on (1.17), (1.19), (1.21) and (1.41) as

$$\frac{dQ}{dW^{(a)}}(X) = \frac{q(X_a, X_b)}{\mu_a(X_a) \, p(a, X_a; b, X_b)} \tag{1.45}$$

$$= \frac{\hat{\varphi}(X_a) \varsigma(X_a) \, p(a, X_a; b, X_b) \, \varphi(X_b)}{\hat{\varphi}(X_a) \varsigma(X_a) \int p(a, X_a; b, z) \, dz \, \varphi(z) \, p(a, X_a; b, X_b)} = \frac{\varphi(X_b)}{\varphi(a, X_a)}$$

and based on (1.18), (1.19), (1.22) and (1.41) as

$$\frac{dQ}{dW^{(b)}}(X) = \frac{q(X_a, X_b)}{\mu_b(X_b) \, p(a, X_a; b, X_b)} \tag{1.46}$$

$$= \frac{\hat{\varphi}(X_a) \varsigma(X_a) \, p(a, X_a; b, X_b) \, \varphi(X_b)}{\int \hat{\varphi}(x) \varsigma(x) \, dx \, p(a, x; b, X_b) \, \varphi(X_b) \, p(a, X_a; b, X_b)} = \frac{\hat{\varphi}(X_a)}{\hat{\varphi}(b, X_b)}$$

Since the Gaussian transition density p is a fundamental solution of the heat equation (1.24) and

$$\triangle \log \kappa = \frac{\triangle \kappa}{\kappa} - (\nabla \log \kappa)^2 \quad \text{for } \kappa \in C^2(\mathbb{R}^d)$$

Itô's formula applied to $\log \varphi$ yields

$$\frac{\varphi(X_b)}{\varphi(a, X_a)} = \exp\{\log \varphi(X_b) - \log \varphi(a, X_a)\}$$

$$= \exp\{\int_a^b \nabla \log \varphi(t, X_t) \, dX_t + \int_a^b (\frac{\partial}{\partial t} + \frac{1}{2} \triangle) \log \varphi(t, X_t) \, dt\} =$$

$$= \exp\{\int_a^b \nabla \log \varphi(t, X_t)\, dX_t - \frac{1}{2} \int_a^b (\nabla \log \varphi(t, X_t))^2 \, dt$$

$$+ \int_a^b \frac{1}{\varphi(t, X_t)} \left(\frac{\partial \varphi(t, X_t)}{\partial t} + \frac{1}{2} \triangle \varphi(t, X_t)\right) dt\}.$$

The third term in the last exponent vanishes since φ satisfies (1.23). Hence the Radon-Nikodym derivative (1.45) turns out to be

$$\frac{dQ}{dW^{(a)}}(X) = \exp\{\int_a^b \frac{\nabla \varphi(t, X_t)}{\varphi(t, X_t)} \, dX_t - \frac{1}{2} \int_a^b \left| \frac{\nabla \varphi(t, X_t)}{\varphi(t, X_t)} \right|^2 \, dt\} \qquad (1.47)$$

which is *Maruyama's density* of the drift

$$\beta_t^{(b)} = \frac{\nabla \varphi(t, X_t)}{\varphi(t, X_t)}, \qquad a \le t \le b. \qquad (1.48)$$

The time-reversal $(\hat{X}_t,\ a \le t \le b,\ Q)$ has the initial distribution μ_b as required by (1.40). In order to determine its drift, we consequently investigate the Radon-Nikodym derivative of Q with respect to $W^{(b)}$ in (1.46). An application of Itô's formula to $\log \hat{\varphi}(a + b - t, \ .\ )$ yields

$$\frac{\hat{\varphi}(a, X_b)}{\hat{\varphi}(b, X_a)} = \exp\{\log \hat{\varphi}(a, X_b) - \log \hat{\varphi}(b, X_a)\}$$

$$= \exp\{\int_a^b \nabla \log \hat{\varphi}(a + b - t, X_t)\, dX_t$$

$$+ \int_a^b \left(\frac{\partial}{\partial t} + \frac{1}{2} \triangle\right) \log \hat{\varphi}(a + b - t, X_t)\, dt\}$$

$$= \exp\{\int_a^b \nabla \log \hat{\varphi}(a + b - t, X_t)\, dX_t - \frac{1}{2} \int_a^b (\nabla \log \hat{\varphi}(a + b - t, X_t))^2 \, dt$$

$$+ \int_a^b \frac{1}{\hat{\varphi}(a + b - t, X_t)} \left(-\frac{\partial \hat{\varphi}(a + b - t, X_t)}{\partial t} + \frac{1}{2} \triangle \hat{\varphi}(a + b - t, X_t)\right) dt\}$$

where the third term in the exponent of the right-hand side vanishes since $\hat{\varphi}$ satisfies (1.24). Hence the Radon-Nikodym derivative in the time-reversed situation takes the form

$$\frac{dQ}{dW^{(b)}}(\hat{X}) = \exp\{\int_a^b \frac{\nabla \hat{\varphi}(a + b - t, X_t)}{\hat{\varphi}(a + b - t, X_t)} \, dX_t \qquad (1.49)$$

$$- \frac{1}{2} \int_a^b \left| \frac{\nabla \hat{\varphi}(a + b - t, X_t)}{\hat{\varphi}(a + b - t, X_t)} \right|^2 \, dt\}$$

which is *Maruyama's density* of the drift

$$\beta_t^{(a)} = \frac{\nabla \hat{\varphi}(a+b-t, X_t)}{\hat{\varphi}(a+b-t, X_t)}, \quad a \leq t \leq b. \tag{1.50}$$

If e.g. the drift vectors (1.48) and (1.50) satisfy the Novikov (1973) condition

$$W^{(r)}[\exp\{\frac{1}{2} \int_a^b | \beta_t^{(r)} |^2 \, dt\}] < \infty, \quad r = a, b \tag{1.51}$$

then the Maruyama-Girsanov drift transformation in terms of the *multiplicative functionals* (1.47) and (1.49), respectively, can be applied. It claims that the Schrödinger process $(X_t, \ a \leq t \leq b, \ Q)$ is represented by the stochastic differential equation

$$dX_t = \frac{\nabla \varphi(t, X_t)}{\varphi(t, X_t)} \, dt + dB_t^{(a)}, \quad a \leq t \leq b$$

where $(B_t^{(a)})_{a \leq t \leq b}$ is the Brownian motion with respect to the probability measure Q defined in (1.44). Moreover, the time-reversed Schrödinger process $(\hat{X}_t, \ a \leq t \leq b, \ Q)$ is represented as

$$dX_t = \frac{\nabla \hat{\varphi}(a+b-t, X_t)}{\hat{\varphi}(a+b-t, X_t)} \, dt + dB_t^{(b)}, \quad a \leq t \leq b$$

where $(B_t^{(b)})_{a \leq t \leq b}$ is the Brownian motion with respect to the probability measure $\hat{Q}$ determined by $d\hat{Q}(X_t) = dQ(X_{a+b-t})$ for $a \leq t \leq b$.

In case of Schrödinger processes corresponding to solutions of Schrödinger equations which are not ground state solutions, see Section 1.4, non-smooth Schrödinger multipliers $\hat{\varphi}$ and φ with *zeros* have to be expected. Consequently, the Schrödinger process $(X_t, \ a \leq t \leq b, \ Q)$ must not cross $\{\hat{\varphi} \varphi = 0\}$ where the drift vectors (1.48) and (1.50) are not well-defined. On the other hand, the Schrödinger system (1.17) & (1.18) represents continuity conditions having their origin in (1.2) and (1.3), respectively. They express that every path leaving at time a must arrive at time b under the measure Q. Such circumstances will be investigated in Chapter 2 under assumptions different from (1.51). As a conclusion we can say that Schrödinger multipliers play their role also in case of $\hat{\varphi}$ and φ with non-vanishing zero sets.

Chapter 2

Diffusions with Singular Drift

Schrödinger equations are shown in Section 2.1 to be equivalent to pairs of adjoint non-linear diffusion equations. Their weak solutions are treated in Section 2.2 as elements of the local space-time Sobolev space $H_{loc}^{1,2}$ which is motivated from the L^2-theory of Schrödinger equations.

Diffusion processes related to Schrödinger equations have in general non-smooth and singular drift as deduced in Section 2.2. They are contained in a wide class of diffusion processes which will be constructed in Section 2.3 by means of a generalized Maruyama-Girsanov drift transformation. Nagasawa's multiplicative functional N_s^t employed as a Radon-Nikodym derivative on the space of continuous paths provides an in general singular drift. The key to deal with non-smooth drift coefficients is a version of Itô's formula in Chapter 4 for continuous space-time functions with first and second order derivatives in the sense of distributions.

Properties of diffusions with singular drift which are conservative, i.e., mass preserving, are deduced in Section 2.3. They enable us in Section 2.4 to answer Schrödinger's question of Section 1.1 in terms of diffusion processes. A sufficient condition for a conservative diffusion with singular drift is obtained from the uniqueness of solutions of Feynman-Kac integral equations with locally integrable potential function. In fact, their solutions are shown in Chapter 3 to correspond to weak solutions of diffusion equations with singular creation and killing treated in Section 2.2 and vice versa.

2.1   Schrödinger equations

In 1931 Schrödinger stated the time-reversibility of natural laws given an observation at two fixed time points, one in the past and one in the future, respectively. As a basic model for a natural law he discussed the case of Brownian motion. In this chapter we consider diffusions with measurable, in general unbounded and only locally integrable creation and killing potential c from Schrödinger's point of view. They can be characterized by the space-time generator $L + c$ where

$$L = \frac{\partial}{\partial s} + \frac{1}{2}\,\triangle + \beta\,\nabla. \tag{2.1}$$

The elliptic differential operator $\triangle$ is either an α-Laplacian

$$\triangle_\alpha = \sum_{i,j=1}^{d} \alpha_{ij}(s,x)\,\frac{\partial^2}{\partial x_i\,\partial x_j}$$

or a Laplacian of α-divergence type

$$\nabla \alpha \nabla = \triangle_\alpha + \sum_{i,j=1}^{d} \frac{\partial \alpha_{ij}}{\partial x_i}\,\frac{\partial}{\partial x_j} \tag{2.2}$$

where the uniformly elliptic and symmetric diffusion matrix α as well as the vector potential β are assumed to possess smooth and bounded coefficients. The Laplacian of α-divergence type is symmetric and hence convenient when considering adjoint diffusion equations. However, it yields the additional drift vector

$$((\alpha \nabla)_j = \sum_{i=1}^{d} \frac{\partial \alpha_{ij}}{\partial x_i})_{1 \le j \le d}. \tag{2.3}$$

Consequently, we prefer the α-Laplacian in connection with stochastic differential equations.

In view of diffusions in Schrödinger-Nagasawa representation, so-called Schrödinger processes as discussed in Section 1.5, we are going to consider the potential c associated with the parabolic differential operator L and the

Schrödinger multiplier $\varphi \in C^{1,2}(\{\varphi > 0\})$ given as

$$c(s, x) = -\frac{L\varphi}{\varphi}(s, x) \quad \text{for} \quad (s, x) \in \{\varphi > 0\}.$$

This chapter intends to establish Schrödinger processes related to Schrödinger equations (2.5) as the real-valued counterparts to solutions Ψ of (2.5). In the present section we want to investigate the related analysis, assuming sufficiently smooth functions Ψ for simplicity. Given the Hamiltonian

$$H = -\frac{1}{2}\triangle + \beta \cdot \nabla + V \tag{2.4}$$

with a real-valued potential V, we can write down the *Schrödinger equation*

$$i\frac{\partial \Psi}{\partial s} = H\Psi. \tag{2.5}$$

Proposition 2.1.1 (Nelson 1966) *Let Ψ be a function of the form*

$$\Psi(s, x) = \exp\{R(s, x) + i\, S(s, x)\}$$

on $D(\Psi) = \{(s, x) \in [a, b] \times \mathbb{R}^d : 0 <| \Psi(s, x) |< \infty\}$ where R, $S \in C^{1,2}(D(\Psi))$.

Then Ψ is a solution of the Schrödinger equation (2.5) on $D(\Psi)$ if and only if the pair of functions (R, S) solves the system of non-linear differential equations

$$V = -\frac{\partial S}{\partial s} + \frac{1}{2}\triangle R + \frac{1}{2}(\sigma \nabla R)^2 - \frac{1}{2}(\sigma \nabla S)^2 - \beta \nabla S \tag{2.6}$$

$$0 = \frac{\partial R}{\partial s} + \frac{1}{2}\triangle S + (\sigma \nabla R)(\sigma \nabla^T S) + \beta \nabla R \tag{2.7}$$

on $D(\Psi)$.

Proof. Since the diffusion matrix α has by assumption a symmetric square root σ, (2.5) can be written on $D(\Psi)$ as

$$i\,\Psi\,\frac{\partial R}{\partial s} - \Psi\,\frac{\partial S}{\partial s} + \Psi\,\frac{1}{2}\,\triangle R + i\,\Psi\,\frac{1}{2}\,\triangle S + i\,\Psi\,(\sigma \nabla R)(\sigma \nabla^T S)$$
$$+ \Psi\,\frac{1}{2}\,(\sigma \nabla R)^2 - \Psi\,\frac{1}{2}\,(\sigma \nabla S)^2 + i\,\beta\,\Psi\,\nabla R - \beta\,\Psi\,\nabla S - \Psi\,V = 0.$$

After dividing the equation by Ψ, its real part yields (2.6) and its imaginary part yields (2.7). These manipulations work on $D(\Psi)$ also in reversed order. $\diamondsuit$

Proposition 2.1.2 (Nagasawa 1989) *Let $(\hat{\varphi}, \varphi)$ be a pair of non-negative functions on $[a, b] \times \mathbb{R}^d$ such that*

$$R = \frac{1}{2} \log(\hat{\varphi}\,\varphi)\,, \quad S = \frac{1}{2} \log(\frac{\varphi}{\hat{\varphi}}) \qquad (2.8)$$

are in $C^{1,2}(D(\hat{\varphi}\,\varphi))$ where $D(\hat{\varphi}\,\varphi) = \{\,(s, x) \in [a, b] \times \mathbb{R}^d : 0 < \hat{\varphi}\varphi < \infty\}$. Let β be a vector potential satisfying the gauge condition $\nabla\beta = 0$. Then

(i)

$$\varphi = \exp\{R + S\} \qquad (2.9)$$

is a solution of

$$\frac{\partial p}{\partial s} + \frac{1}{2}\nabla\alpha\nabla p + \beta\,\nabla p + c_\varphi\,p = 0 \quad on \ \ D(\hat{\varphi}\,\varphi) \qquad (2.10)$$

if and only if R and S in (2.8) satisfy on $D(\hat{\varphi}\,\varphi)$

$$
\begin{aligned}
c_\varphi \ = \ & \{-\frac{\partial S}{\partial s} + \frac{1}{2}\nabla\alpha\nabla R + \frac{1}{2}(\sigma\,\nabla R)^2 - \frac{1}{2}(\sigma\,\nabla S)^2 - \beta\,\nabla S\} \quad (2.11) \\
& - \{\frac{\partial R}{\partial s} + \frac{1}{2}\nabla\alpha\nabla S + (\sigma\,\nabla R)\,(\sigma\,\nabla^T S) + \beta\,\nabla R\} \\
& - \{\nabla\alpha\nabla R + (\sigma\,\nabla R)^2\}\,.
\end{aligned}
$$

(ii)

$$\hat{\varphi} = \exp\{R - S\} \qquad (2.12)$$

is a solution of

$$-\frac{\partial p^*}{\partial s} + \frac{1}{2}\nabla\alpha\nabla p^* - \beta\,\nabla p^* + c_{\hat{\varphi}}\,p^* = 0 \quad on \ \ D(\hat{\varphi}\,\varphi) \qquad (2.13)$$

if and only if R and S in (2.8) satisfy on $D(\hat{\varphi}\,\varphi)$

$$
\begin{aligned}
c_{\hat{\varphi}} \ = \ & \{-\frac{\partial S}{\partial s} + \frac{1}{2}\nabla\alpha\nabla R + \frac{1}{2}(\sigma\,\nabla R)^2 - \frac{1}{2}(\sigma\,\nabla S)^2 - \beta\,\nabla S\} \quad (2.14) \\
& + \{\frac{\partial R}{\partial s} + \frac{1}{2}\nabla\alpha\nabla S + (\sigma\,\nabla R)\,(\sigma\,\nabla^T S) + \beta\,\nabla R\} \\
& - \{\nabla\alpha\nabla R + (\sigma\,\nabla R)^2\}\,.
\end{aligned}
$$

Proof. We plug (2.9) in (2.10) and (2.12) in (2.13). After dividing (2.10) and (2.13) by $\hat{\varphi}$ and φ, respectively, the potentials c_φ and $c_{\hat{\varphi}}$ can be separated. $\Diamond$

Theorem 2.1.1 (Nagasawa 1989) *Let us consider any differential operator consisting of* $\nabla \alpha \nabla$, *a vector potential* β *with* $\nabla \beta = 0$ *and a real-valued potential function* V.

(*i*) *Let* $(\varphi, \hat{\varphi})$ *be a solution of the system (2.10) & (2.13) of* non-linear interacting diffusion equations *on* $D(\hat{\varphi}\,\varphi)$ *with* $c_{\hat{\varphi}} = c_{\varphi} = c_{\hat{\varphi}\,\varphi}$. *Then the wave function*

$$\Psi = e^{R + i\,S} \tag{2.15}$$

in terms of R *and* S *in (2.8) is a solution of the Schrödinger equation (2.5) on* $D(\hat{\varphi}\,\varphi)$ *with the potential*

$$V_{\Psi} = -c_{\hat{\varphi}\,\varphi} - 2\frac{\partial S}{\partial s} - (\sigma\,\nabla S)^2 - 2\beta\,\nabla S. \tag{2.16}$$

(*ii*) *Let* Ψ *in (2.15) determined by* R, $S \in C^{1,2}(D(\Psi))$ *be a solution of the Schrödinger equation (2.5). Then, the pair of functions* $(\varphi, \hat{\varphi})$ *given in (2.9) and (2.12) is a solution of the system (2.10) & (2.13) of* non-linear interacting diffusion equations *on* $D(\Psi)$ *where the creation and killing potential consists of*

$$c_{\hat{\varphi}\,\varphi} = -V_{\Psi} - 2\frac{\partial S}{\partial s} - (\sigma\,\nabla S)^2 - 2\beta\,\nabla S. \tag{2.17}$$

Proof. (*i*) Because of (2.11) and (2.14), $c_{\varphi} = c_{\hat{\varphi}} = c_{\hat{\varphi}\,\varphi}$ yields (2.7) and (2.6) for the potential

$$V_{\Psi} = c_{\hat{\varphi}\,\varphi} + \nabla\alpha\nabla R + (\sigma\,\nabla R)^2. \tag{2.18}$$

Hence the converse part of Proposition 2.1.1 claims that Ψ in (2.15) satisfies (2.5) with the potential in (2.18). Thus (2.6) yields (2.16) for V_{Ψ} in (2.18).

(*ii*) Following Proposition 2.1.1, (2.6) and (2.7) hold. Proposition 2.1.2 shows that c_{φ} and $c_{\hat{\varphi}}$ differ only in the sign of the second member of the right-hand side - but this term vanishes because of (2.7). According to (2.6), the first members of the right-hand sides of (2.11) and (2.14) equal V_{Ψ}; hence (2.17) follows. $\Diamond$

Theorem 2.1.1 claims that the Schrödinger equation (2.5) is equivalent to the pair of adjoint non-linear interacting diffusion equations in (2.10) and (2.13) with $c_{\varphi} = c_{\hat{\varphi}} = c_{\hat{\varphi}\,\varphi}$. Some consequences to potential functions will be provided by Section 6.4. Among others, Blanchard-Golin (1987), Carlen (1984), Carmona

(1985), Nelson (1967), Zheng-Meyer (1984) and Zheng (1985) investigate diffusion processes related to solutions Ψ of Schrödinger equations (2.5). Carmona (1985) gives a martingale approach in the spirit of Stroock-Varadhan (1969, 1979). He emploies the change of variable (2.9) in case of $\alpha = 1$, $\beta = 0$ and points out the relation between (2.10) and (2.17).

Let us discuss relevant regularity assumptions under which Schrödinger processes in Section 1.5 related to Schrödinger equations (2.5) have to be constructed. Simon (1982) considers the operator H in (2.4) as a perturbed free energy Hamiltonian with a time-independent potential which defines a multiplication operator of a Kato class function V. He shows that for any Cauchy initial data $\Psi_a \in L^2(I\!\!R^d, \lambda^d)$ at time a, λ^d the Lebesque measure on $I\!\!R^d$, the unique solution of (2.5) in the $L^2(I\!\!R^d, \lambda^d)$-sense is given by

$$e^{-i\,H\,(s-a)}\,\Psi_a \tag{2.19}$$

where $\{e^{-i\,H\,(s-a)}, a \leq s \leq b\}$ is the canonical unitary group on $L^2(I\!\!R^d, \lambda^d)$. If Ψ_a is in the domain of the square root of the Hamiltonian H, Carmona (1985) finds jointly continuous versions Ψ of (2.19) and its time-derivative $\frac{\partial \Psi}{\partial s}$, respectively, as well as $\nabla \Psi \in L^2_{loc}$ and $\triangle \Psi \in L^1_{loc}$ on $[a, b] \times I\!\!R^d$ which replace the gradient and the Laplacian of (2.19) in the sense of distributions, respectively. Considering an analytic version of the logarithm in a simple connected subset $I\!\!D$ of the complex plane,

$$R(s, x) = \text{Re log } \Psi(s, x), \qquad S(s, x) = \text{Im log } \Psi(s, x)$$

are well-defined for $(s, x) \in \Psi^{-1}(I\!\!D)$. Carmona (1985) determines the space-time trajectories which have the marginal distribution density $\overline{\Psi}_t \Psi_t$ at any time $t \in [a, b]$ and which do not leave the set $\Psi^{-1}(I\!\!D)$ almost surely. Referring to (2.9) and (2.12) we notice that the distribution density $\overline{\Psi}_t \Psi_t$ equals the distribution density $\hat{\varphi}\,\varphi(t, \,.\,)$ of a Schrödinger process in (1.38). The function φ in (2.9) inherits the regularity properties of (2.19) and it satisfies (2.10) with (2.11) on $\Psi^{-1}(I\!\!D)$ in the sense of distributions.

We are going to define a wide class of *Schrödinger multipliers* Φ for which we are able to construct diffusion processes with singular drift $\alpha \nabla \log \Phi$. They will yield in Section 2.4 Schrödinger processes related to Schrödinger equations (2.5) treated in the standard $L^2(I\!\!R^d, \lambda^d)$-context. Our strategy is to extend Nagasawa's multiplicative functional $(N^t_s)_{a \leq s \leq t \leq b}$ to non-negative, jointly continuous functions Φ in the local space-time Sobolev space $H_{loc}^{1,2}$ of Definition 2.2.1. The key representation of $(N_s^t)_{a\leq s\leq t\leq b}$ given in Section 2.3 is obtained by a version of Itô's formula for non-smooth function deduced in Chapter 4. A generalized version the Maruyama-Girsanov drift transformation claims in Section 2.3 that the process

$$((t, X_t), \ a \leq t \leq b, \ Q_{(s,x)}, \ (s,x) \in \{\Phi > 0\})$$

constructed by means of the multiplicative functional $(N_s^t)_{a\leq s\leq t\leq b}$ possesses the diffusion matrix α and the in general singular drift

$$\beta + \alpha \, \frac{\nabla^T \Phi}{\Phi}.$$

Its space-time trajectories $(t, X_t)_{s\leq t\leq b}$ do not hit the set of zeros and singularities of Φ, $Q_{(s,x)}$-almost surely. Section 2.4 is devoted to Schrödinger processes and their time-reversal nature.

2.2 Non-smooth Schrödinger multipliers

For any open subset D of $(a,b) \times \mathbb{R}^d$, $-\infty < a < b < \infty$, there exists an increasing sequence $(D^{(n)})_{n\in\mathbb{N}}$ of open sets with compact closure $\overline{D^{(n)}}$ contained in D such that

$$D = \bigcup_{n\in\mathbb{N}} D^{(n)}. \tag{2.20}$$

We set $\mathcal{N} = \{\nu = (\nu_s, \nu_1, \ldots, \nu_d) : 0 \leq \nu_s \leq 1, \ 0 \leq \nu_i \leq 2 \text{ for } 1 \leq i \leq d, \text{ where } \nu_s + \nu_1 + \ldots + \nu_d \leq 3\}$ and consider

$$C_{loc}^{1,2}(D) = \{F : D \to \mathbb{R}; \ F \text{ has continuous first order derivative in } s \text{ as}$$
$$\text{well as continuous first and second order derivatives in } x$$
$$\text{for } (s,x) \in D \text{ such that}$$

$$\| F \|_{D^{(n)}} = (\sum_{\nu\in\mathcal{N}} \int_{D^{(n)}} | D^\nu F(s,x) |^2 \ ds\,dx)^{\frac{1}{2}} < \infty \text{ for any } n \in \mathbb{N}\}$$

where $D^\nu = D_s^{\nu_s} D_1^{\nu_1} \cdot \ldots \cdot D_d^{\nu_d}$. $\| F \|_D = \sum_{n=1}^\infty 2^{-n} (\| F \|_{D^{(n)}} \wedge 1)$ is well-defined for $F \in C_{loc}^{1,2}(D)$ and, by convention, $F_1, F_2 \in C_{loc}^{1,2}(D)$ are identified if $\| F_1 - F_2 \|_{D^{(n)}} = 0$ for every $n \in \mathbb{N}$.

Definition 2.2.1 $H_{loc}^{1,2}(D)$ is the completion of $(C_{loc}^{1,2}(D), \| \cdot \|_D)$, i.e., a local space-time *Sobolev space.*

With $C_{comp}^{\infty}(D)$ as the space of test functions, any $F \in H_{loc}^{1,2}(D)$ has derivatives in the sense of distributions $D^{\nu} F \in L_{loc}^2(D)$, $\nu \in \mathcal{N}$, as adopted from Friedman (1964)'s chap. X, also concerning the notation.

Definition 2.2.2 A function $\Phi : [a, b] \times \mathbb{R}^d \to [0, \infty)$ is said to be a *Schrödinger multiplier* if Φ is jointly continuous on $\{\Phi > 0\} \cap \{| D^{\nu}\Phi |< \infty, \ \nu \in \mathcal{N}\}$ as well as $\Phi \in H_{loc}^{1,2}(D(\Phi))$ for $D(\Phi) = \{\Phi > 0\} \cap \{| D^{\nu}\Phi |< \infty, \ \nu \in \mathcal{N}\} \cap (a, b) \times \mathbb{R}^d$ which is assumed to be an open subset of $\mathbb{R}^{d+1}$. We will write $\nabla = (D_1, \dots, D_d)$ and $\triangle_{\alpha} = \sum_{i,j=1}^d \alpha_{ij} D_i D_j$ for the gradient and the α-Laplacian in the sense of distributions, respectively.

Lemma 2.2.1 (lemma 3.2 in Nagasawa 1989) *Let* φ *and* p *be functions in* $C^{1,2}(\{\varphi > 0\})$ *and let* q *be defined as*

$$q(s, x) = \frac{p(s,x)}{\varphi(s,x)} \quad \text{for} \ (s,x) \in \{\varphi > 0\}.$$

Then

$$Lq + \frac{\nabla\varphi}{\varphi}\nabla q = \frac{1}{\varphi}(Lp + cp) - \frac{p}{\varphi^2}(L\varphi + c\varphi) \quad \text{on} \ \{\varphi > 0\}$$

where L *is the parabolic differential operator in (2.1) and* c *is an arbitrary function on* $\{\varphi > 0\}$.

Proof. This simple verification is routine. ◇

Lemma 2.2.2 *Let* p *be a jointly continuous weak solution of*

$$(L + c)\, p = 0 \quad \text{on} \ D(\Phi) \tag{2.21}$$

where the so-called weak reference potential *c associated with L in (2.1) and Φ in Definition 2.2.2 is defined as*

$$c = -\frac{L\Phi}{\Phi} \quad \text{on} \ D(\Phi) \tag{2.22}$$

and set to be zero elsewhere. Then

$$q(s, x) = \frac{p(s,x)}{\Phi(s,x)} \quad \text{for} \ (s,x) \in D(\Phi) \tag{2.23}$$

is a jointly continuous weak solution of

$$L\,q + \frac{\nabla \Phi}{\Phi}\,\alpha\,\nabla^T q \;=\; 0 \quad on \;\; D(\Phi). \tag{2.24}$$

In other words, a weak solution p of the diffusion equation (2.21) with creation and killing c in (2.22) provides a weak solution q of the diffusion equation (2.24) with the in general singular drift operator

$$\frac{\nabla \Phi}{\Phi}\,\alpha\,\nabla^T.$$

Proof. For any fixed $\gamma \in C^\infty_{comp}(D(\Phi))$, $K = \operatorname{supp}\gamma$, we can find sequences $(\Phi_n)_{n\in\mathbb{N}}$ and $(p_m)_{m\in\mathbb{N}}$ in $C^{1,2}(K)$ which approximate Φ in the $H^{1,2}_{loc}$-sense and p uniformly on K, respectively. There exist constants $M(K)$ and $\delta > 0$ such that $\max_K |p_m| \leq M(K)$ for all $m \geq m(K)$ and $\min_K \Phi_n \wedge \min_K \Phi \geq \delta$ for all $n \geq n(K,\delta)$. We set

$$q_n^m(s,x) \;=\; \frac{p_m(s,x)}{\Phi_n(s,x)}$$

for $(s,x) \in K$, $n \geq n(K,\delta)$, $m \geq m(K)$ and verify, using Lemma 2.2.1, that

$$\int_K [L^*\gamma - \nabla(\frac{\gamma}{\Phi_n}\nabla \Phi_n\,\alpha)]\,q_n^m \tag{2.25}$$

$$= \int_K [L^*(\frac{\gamma}{\Phi_n}) + c\,\frac{\gamma}{\Phi_n}]\,p_m - \int_K \frac{\gamma\,p_m}{\Phi_n^2}\,[L\,\Phi_n + c\,\Phi_n]$$

where L^* denotes the adjoint of L given in (2.1). Since p is a weak solution of (2.21), we can estimate the first member of the right-hand side of (2.25) by

$$\max_K |\,p_m - p\,|\,\|\,L^*(\frac{\gamma}{\Phi_n}) + c\,\frac{\gamma}{\Phi_n}\,\|_{L^1(K)}$$

which vanishes as m tends to infinity for $n \geq n(K,\delta)$. Because of (2.22), the absolute value of the second member of the right-hand side of (2.25) is dominated by

$$\frac{M(K)\,|\,K\,|^{\frac{1}{2}}}{\delta^2}\,\max_K |\,\gamma\,|\,[\,\|\,L\,\Phi_n - L\Phi\,\|_{L^2(K)}$$

$$+ \frac{1}{\delta^2}\,\|\,L\Phi\,\|_{L^2(K)}\,\max_K |\,\Phi_n^2 - \Phi^2\,| + \frac{1}{\delta}\,\|\,L\Phi\,\|_{L^2(K)}\,\max_K |\,\Phi_n - \Phi\,|\,]$$

which vanishes as n tends to infinity for $m \geq m(K)$. Furthermore, we have

$$L^1_{loc}(D(\Phi)) - \lim_{n\to\infty} \nabla(\frac{\gamma}{\Phi_n}\nabla \Phi_n\,\alpha) \;=\; \vartheta(\gamma,\Phi) \tag{2.26}$$

for

$$\vartheta(\gamma, \Phi) = \frac{\gamma}{\Phi} \, \triangle_\alpha \Phi - \gamma \, \frac{\nabla \Phi}{\Phi} \, \alpha \, \frac{\nabla^T \Phi}{\Phi} + \frac{\gamma}{\Phi} \sum_{i,j=1}^{d} \frac{\partial \alpha_{ij}}{\partial x_j} \, D_i \, \Phi + \frac{\nabla \Phi}{\Phi} \, \alpha \, \nabla^T \gamma$$

which is in $L^1_{loc}(D(\Phi))$. As a consequence, the assertion which means

$$\int_K [L^* \gamma - \vartheta(\gamma, \Phi)] \, q \; = \; 0 \tag{2.27}$$

follows from (2.25) and (2.26) as m and n tend to infinity, writing (2.27) as

$$\int_K L^* \gamma \, (q - q_n^m)$$
$$+ \; \int_K [\nabla(\frac{\gamma}{\Phi_n} \nabla \Phi_n \, \alpha) \, q_n^m - \vartheta(\gamma, \Phi) \, q] + \int_K [L^* \gamma - \nabla(\frac{\gamma}{\Phi_n} \nabla \Phi_n \, \alpha)] \, q_n^m.$$

$\diamondsuit$

In view of an application of the generalized version of Itô's formula in Section 4.3, we formulate

Lemma 2.2.3 *A Schrödinger multiplier Φ in Definition 2.2.2 satisfies*

$$\int_D \log \Phi \, (-\frac{\partial}{\partial s} \gamma) \; = \; \int_D \frac{1}{\Phi} \, D_s \Phi \, \gamma \tag{2.28}$$

$$\int_D \log \Phi \, (-\frac{\partial}{\partial x_i} \gamma) \; = \; \int_D \frac{1}{\Phi} \, D_i \Phi \, \gamma \tag{2.29}$$

$$\int_D \log \Phi \, (\frac{\partial^2}{\partial x_i \, \partial x_j} \gamma) \; = \; \int_D (\frac{1}{\Phi} \, D_i D_j \Phi - \frac{1}{\Phi^2} \, D_i \Phi \, D_j \Phi) \, \gamma \tag{2.30}$$

for $1 \leq i, j \leq d$ and test functions $\gamma \in C^\infty_{comp}(D(\Phi))$.

Proof. Let $\gamma \in C^\infty_{comp}(D(\Phi))$ be fixed. Set $K = \operatorname{supp} \gamma$ and $\delta = \frac{1}{2} \min_K \Phi > 0$. Definition 2.2.1 provides a sequence $(\Phi_n)_{n \in \mathbb{N}} \subset C^{1,2}(K)$ which approximates Φ in the $H^{1,2}_{loc}$-sense as well as uniformly on K. Hence there exists $n(K, \delta) \in \mathbb{N}$ such that $\min_K \Phi_n \geq \delta$ for all $n \geq n(k, \delta)$. We receive (2.28) by

$$| \int_D (\log \Phi \, (-\frac{\partial}{\partial s} \gamma) - \frac{1}{\Phi} \, D_s \Phi \, \gamma) |$$
$$\leq \; \{\int_K (-\frac{\partial}{\partial s} \gamma)^2\}^{\frac{1}{2}} \, \{\int_K (\log \Phi - \log \Phi_n)^2\}^{\frac{1}{2}} +$$

$$+ \frac{\mid K \mid^{\frac{1}{2}} \max_K \mid \gamma \mid}{\delta} \{ \int_K (\frac{\partial}{\partial s} \Phi_n - D_s \Phi)^2 \}^{\frac{1}{2}}$$

$$+ \frac{\max_K \mid \gamma \mid}{\delta^2} \{ \int_K (D_s \Phi)^2 \}^{\frac{1}{2}} \{ \int_K (\Phi - \Phi_n)^2 \}^{\frac{1}{2}}$$

since the right-hand side vanishes as n tends to infinity. (2.29) and (2.30) are shown similarly. $\Diamond$

Remark 2.2.1 Lemma 2.2.3 shows that $\log \Phi$ is in general *not* an element of $H_{loc}^{1,2}(D(\Phi))$. In fact,

$$D_i D_j \log \Phi = \frac{1}{\Phi} D_i D_j \Phi - \frac{1}{\Phi^2} D_i \Phi \, D_j \Phi$$

is in general not locally square-integrable. However, it is locally integrable and approximated by $(D_i D_j \log \Phi_n)_{n \in \mathbb{N}}$ in the $L_{loc}^1(D(\Phi))$-sense.

2.3 Singular transformation of diffusions

Itô's formula for non-smooth functions in Section 4.3 is formulated in terms of diffusion processes which are considered as path-valued random variables $(\xi_{s,t}(x)(\omega))_{s \leq t \leq b}$, $\omega \in \Omega$, $(s, x) \in [a, b] \times \mathbb{R}^d$, defined on a probability space $(\Omega, \mathcal{F}, (\mathcal{F}_t)_{a \leq t \leq b}, P)$. In this chapter however, diffusion processes are conveniently considered as measures on a space of continuous path. Hence we state

Definition 2.3.1 The diffusion process $(\xi_{s,t}(x))_{s \leq t \leq b}$ in (4.10) on $C([a, b], \mathbb{R}^d)$ associated with the parabolic differential operator L in (2.1) is considered as a family of probability measures $P_{(s,x)}$, $(s, x) \in [a, b] \times \mathbb{R}^d$, which are by Stroock-Varadhan (1979)'s chap. 5 uniquely determined by

$$P_{(s,x)}(X_{t_1} \in \Gamma_1, \ldots, X_{t_n} \in \Gamma_n) = P(\xi_{s,t_1}(x) \in \Gamma_1, \ldots, \xi_{s,t_n}(x) \in \Gamma_n)$$

for $s \leq t_1 < \ldots < t_n \leq b$ and $\Gamma_1, \ldots, \Gamma_n \in \sigma_{Borel}(\mathbb{R}^d)$ where $X \in C([a, b], \mathbb{R}^d)$.

Definition 2.3.2 Nagasawa's multiplicative functional $(N_s^t)_{s \leq t \leq b}$, $a \leq s \leq b$, is defined on the space of continuous space-time paths $(t, X_t)_{a \leq t \leq b}$ as

$$N_s^t = \exp\{ \int_s^t c(u, X_u) \, du \} \frac{\Phi(t, X_t)}{\Phi(s, X_s)} 1_{\{t < \tau_s\}} \qquad (2.31)$$

for Schrödinger multipliers Φ in Definition 2.2.2 and the weak reference potential c associated with L and Φ in (2.22) where

$$\tau_s = \inf\{u \in [s,b) : (u, X_u) \notin D(\Phi)\}. \tag{2.32}$$

Itô's formula for non-smooth functions in Theorem 4.3.1 allows us to investigate $(N_s^t)_{s \le t \le b}$ in (2.31) with Schrödinger multipliers Φ from Definition 2.2.2 for the purpose of drift transformation in Corollary 2.3.2.

Corollary 2.3.1 (representation theorem) *Let $(D^{(n)})_{n \in I\!N}$ be a sequence of open sets which approximates $D(\Phi)$ according to (2.20).*

Then
$$\tau_s^{(n)} = \inf\{u \in [s,b) : (u, X_u) \notin D^{(n)}\} \tag{2.33}$$

satisfies
$$\lim_{n \nearrow \infty} \tau_s^{(n)} = \tau_s \quad on \ \{\tau_s < \infty\}, \quad P_{(s,x)}\text{-}a.s.$$

for $(s,x) \in D(\Phi)$ and defines well the limit
$$\Gamma_s^t(\tfrac{\nabla\Phi}{\Phi}\sigma(\,.\,,X_.\,))$$
$$= P_{(s,x)} - \lim_{n \nearrow \infty} 1_{\{\int_s^t \|\frac{\nabla\Phi}{\Phi}\sigma(u,X_u)\|^2\,du < \infty\}} \int_s^{t \wedge \tau_s^{(n)}} \tfrac{\nabla\Phi}{\Phi}\sigma(u, X_u)\,dX_u$$

for $t \in [s,b]$ where $P_{(s,x)} - \lim$ denotes convergence in probability and $\nabla\Phi/\Phi$ is defined to vanish outside of $D(\Phi)$.

Moreover, $(N_s^t)_{s \le t \le b}$ in Definition 2.3.2 can be represented as
$$N_s^t = 1_{\{t < \tau_s\}} \exp\{\Gamma_s^t(\tfrac{\nabla\Phi}{\Phi}\sigma(\,.\,,X_.\,)) - \frac{1}{2}\int_s^t \tfrac{\nabla\Phi}{\Phi}\alpha\tfrac{\nabla^T\Phi}{\Phi}(u, X_u)\,du\} \tag{2.34}$$

$P_{(s,x)}$-*a.s. for λ^d-a.a. $x \in I\!R^d$ with $(s,x) \in D(\Phi)$.*

If $(N_s^t)_{s \le t \le b}$ is in addition continuous along $(\tau_s^{(n)})_{n \in I\!N}$ at time τ_s on the set $\{\tau_s < \infty\}$, $P_{(s,x)}$-a.s. for λ^d-a.a. $x \in I\!R^d$ with $(s,x) \in D(\Phi)$, then
$$N_s^t = \exp\{\Gamma_s^t(\tfrac{\nabla\Phi}{\Phi}\sigma(\,.\,,X_.\,)) - \frac{1}{2}\int_s^t \tfrac{\nabla\Phi}{\Phi}\alpha\tfrac{\nabla^T\Phi}{\Phi}(u, X_u)\,du\} \tag{2.35}$$

and
$$\int_s^{\tau_s} \|\tfrac{\nabla\Phi}{\Phi}\sigma(u, X_u)\|^2\,du = \infty \quad on \ \{\tau_s < \infty\} \tag{2.36}$$

where both hold $P_{(s,x)}$-a.s. for λ^d-a.a $x \in I\!R^d$ with $(s,x) \in D(\Phi)$.

Proof. $(N_s^t)_{s \leq t \leq b}$ in Definition 2.3.2 is a $\sigma(X_v, \ s \leq v \leq t)$-adapted, non-negative multiplicative functional on $C([a,b], I\!\!R^d)$ which is $P_{(s,x)}$-a.s. continuous, except at time τ_s where it may have a jump down to zero. For a discussion of multiplicative functionals we refer to Dynkin (1965)'s vol. 1, chap. VI and IX.

For a verification of (2.34), Lemma 2.2.3 provides $D_s \log \Phi \in L^2_{loc}(D(\Phi))$, $D_i \log \Phi \in L^2_{loc}(D(\Phi))$ and $D_i D_j \log \Phi \in L^1_{loc}(D(\Phi))$ for $1 \leq i,j \leq d$. Hence Itô's formula for non-smooth functions in Theorem 4.3.1 can be applied to $\log \Phi$ where Φ in Definition 2.2.2 is generated along the space-time diffusion process

$$(t, \xi_{s,t}(x))_{s \leq t \leq b}$$

associated with L in (2.1) and stopped by

$$\tau_{(s,x)}^{(n)} = \inf\{u \in [s,b) : (u, \xi_{s,u}(x)) \notin D^{(n)}\}.$$

We receive

$$\log \Phi(t \wedge \tau_{(s,x)}^{(n)}, \xi_{s,t\wedge\tau_{(s,x)}^{(n)}}(x)) - \log \Phi(s,x)$$

$$= \int_s^{t\wedge\tau_{(s,x)}^{(n)}} \{\frac{D_s \Phi}{\Phi}(u, \xi_{s,u}(x)) + \sum_{i=1}^d \frac{D_i \Phi}{\Phi}(u, \xi_{s,u}(x)) \beta^i(u, \xi_{s,u}(x))$$

$$+ \frac{1}{2} \sum_{i,j=1}^d (\frac{D_i D_j \Phi}{\Phi} - \frac{D_i \Phi D_j \Phi}{\Phi^2})(u, \xi_{s,u}(x))$$

$$\cdot \sum_{k=1}^d \sigma^{ik}(u, \xi_{s,u}(x)) \sigma^{jk}(u, \xi_{s,u}(x))\} \, du$$

$$+ \int_s^{t\wedge\tau_{(s,x)}^{(n)}} \sum_{j=1}^d \frac{D_j \Phi}{\Phi}(u, \xi_{s,u}(x)) \sum_{k=1}^d \sigma^{jk}(u, \xi_{s,u}(x)) \, dB_u^k$$

for $t \in [s,b]$, $\lambda^d \otimes P$-a.s. on $\{x \in I\!\!R^d : (s,x) \in D(\Phi)\} \times \Omega$, where $(B_t)_{s \leq t \leq b}$ is the d-dimensional Brownian motion w.r.t. P. This can be reformulated according to Definition 2.3.1 and in terms of (2.22) as

$$\exp\{\int_s^{t\wedge\tau_s^{(n)}} c(u, X_u) \, du\} \frac{\Phi(t \wedge \tau_s^{(n)}, X_{t\wedge\tau_s^{(n)}})}{\Phi(s,x)} \qquad (2.37)$$

$$= \exp\{\int_s^{t\wedge\tau_s^{(n)}} \frac{\nabla\Phi}{\Phi} \sigma(u, X_u) \, dB_u - \frac{1}{2} \int_s^{t\wedge\tau_s^{(n)}} \frac{\nabla\Phi}{\Phi} \alpha \frac{\nabla^T\Phi}{\Phi}(u, X_u) \, du\}$$

for $s \leq t \leq b$, $P_{(s,x)}$-a.s. for λ^d-a.a. $x \in I\!\!R^d$ with $(s,x) \in D(\Phi)$. Since the right-hand side of (2.37) is just a stopped Maruyama density, i.e., a stopped exponential martingale as considered in Liptser-Shiryayev (1977)'s chap. VI, Fatou's lemma yields

$$P_{(s,x)}[N_s^t] \le 1, \quad s \le t \le b \qquad (2.38)$$

for $s \in [a, b]$ and λ^d-a.a. $x \in \mathbb{R}^d$ with $(s, x) \in D(\Phi)$. Denoting

$$\chi_t = 1_{\{\int_s^t \|\frac{\nabla \Phi}{\Phi} \sigma(u, X_u)\|^2 \, du < \infty\}}$$

we obtain

$$
\begin{aligned}
N_s^t &= 1_{\{t < \tau_s\}} \, P_{(s,x)} - \lim_{n \nearrow \infty} \chi_t \, N_s^{t \wedge \tau_s^{(n)}} \\
&= 1_{\{t < \tau_s\}} \, \exp\{P_{(s,x)} - \lim_{n \nearrow \infty} \chi_t \int_s^{t \wedge \tau_s^{(n)}} \frac{\nabla \Phi}{\Phi} \sigma(u, X_u) \, dB_u \\
&\qquad\qquad - \frac{1}{2} \int_s^t \| \frac{\nabla \Phi}{\Phi} \sigma(u, X_u) \|^2 \, du\}
\end{aligned}
$$

$P_{(s,x)}$-a.s. for λ^d-a.a. $x \in \mathbb{R}^d$ with $(s, x) \in D(\Phi)$, which is (2.34) as a consequence of Liptser-Shiryayev (1977)'s chap. IV.

If $\lim_{n \to \infty} N_s^{t \wedge \tau_s^{(n)}} = N_s^t$ for $t \in [s, b]$, $P_{(s,x)}$-a.s., then

$$N_s^t = \chi_t \, \exp\{\Gamma_s^{t \wedge \tau_s}(\frac{\nabla \Phi}{\Phi} \sigma(\,.\,, X_{\,.\,})) - \frac{1}{2} \int_s^{t \wedge \tau_s} \| \frac{\nabla \Phi}{\Phi} \sigma(u, X_u) \|^2 \, du\} \quad (2.39)$$

as well as

$$\chi_{\tau_s} \, \exp\{\Gamma_s^{\tau_s}(\frac{\nabla \Phi}{\Phi} \sigma(\,.\,, X_{\,.\,})) - \frac{1}{2} \int_s^{\tau_s} \| \frac{\nabla \Phi}{\Phi} \sigma(u, X_u) \|^2 \, du\} = 0 \qquad (2.40)$$

on $\{\tau_s < \infty\}$, $P_{(s,x)}$-a.s. for λ^d-a.a. $x \in \mathbb{R}^d$ with $(s, x) \in D(\Phi)$. If

$$\mathcal{A} = \{\tau_s < \infty\} \cap \{\int_s^{\tau_s} \| \frac{\nabla \Phi}{\Phi} \sigma(u, X_u) \|^2 \, du < \infty\}$$

has strictly positive probability, then also

$$\mathcal{A} \cap \{\sup_{n \in \mathbb{N}} | \int_s^{\tau_s^{(n)}} \frac{\nabla \Phi}{\Phi} \sigma(u, X_u) \, dB_u | < \infty\}$$

by lemma 4.7 in Liptser-Shiryayev (1977). In order to avoid a contradiction with (2.40), we arrive at (2.36) which completes together with (2.39) the proof of (2.35). ◊

Corollary 2.3.2 *If a Schrödinger multiplier* Φ *in Definition 2.2.2 and its reference potential c in (2.22) satisfy (3.31), i.e.,*

$$P_{(s,x)}[\exp\{\int_s^t c(u, X_u)\, du\}\, \Phi(t, X_t); t < \tau_s] < \infty$$

and (3.32), i.e.,

$$\int_s^t dv\, P_{(s,x)}[\exp\{\int_s^v c(u, X_u)\, du\}\, c^+(v, X_v)\, \Phi(v, X_v); v < \tau_s] < \infty$$

or (3.30), i.e.,

$$P_{(s,x)}[\Phi(t, X_t); t < \tau_s] < \infty$$

and (3.33), i.e.,

$$\int_s^t dv\, P_{(s,x)}[\exp\{\int_s^v |\, c(u, X_u)\, |\, du\}\, |\, c(v, X_v)\, \Phi(v, X_v)\, |; v < \tau_s] < \infty$$

for $(s, x) \in D(\Phi)$, $a \leq s \leq t \leq b$, *with* τ_s *in (2.32) and* $D(\Phi)$ *in Definition 2.2.2, then*

$$P_{(s,x)}[N_s^t] = 1, \quad s \leq t \leq b \tag{2.41}$$

for $a \leq s \leq b$, λ^d*-a.e. on* $\{x \in \mathbb{R}^d : (s, x) \in D(\Phi)\}$. *As a consequence, there exists a conservative transformed space-time diffusion process*

$$((t, X_t),\, \sigma((v, X_v),\, s \leq v \leq t),\, Q_{(s,x)},\, (s, x) \in D(\Phi),\, s \leq t \leq b) \tag{2.42}$$

determined by the transition probability

$$q(s, x; t, f) = Q_{(s,x)}[f(t, X_t)] = P_{(s,x)}[f(t, X_t)\, N_s^t] \tag{2.43}$$

for bounded and measurable functions f on $D(\Phi)$. *Its trajectories are* $Q_{(s,x)}$*-a.s. continuous and they do not hit the set* $D(\Phi)^c$ *of zeros and singularities of* Φ, $Q_{(s,x)}$*-a.s. Moreover, the representation (2.35) as well as the property (2.36) hold.*

Proof. Corollary 3.4.2 provides (2.41) from which the existence of (2.42) follows by means of the main theorem in H. Kunita-T. Watanabe (1963). In case of $(s, x) \in D(\Phi)$

$$Q_{(s,x)}[1_{D(\Phi)}(t, X_t); t < \zeta] = P_{(s,x)}[1_{D(\Phi)}(t, X_t)\, N_s^t] = P_{(s,x)}[N_s^t] = 1$$

holds for $s \leq t \leq b$, where ζ denotes the life time of (2.42). Hence the $Q_{(s,x)}$-almost certainly continuous paths of the transformed process do not hit $Q_{(s,x)}$-a.s. the set of zeros and singularities of Φ which is the complementary set of $D(\Phi)$ denoted by $D(\Phi)^c$. $\Diamond$

The transformed processes $(X_t,\ Q_{(s,x)},\ (s,x) \in D(\Phi),\ s \leq t \leq b)$ in Corollary 2.3.2 will be the crucial ingredient in the construction of Schrödinger processes given in Section 2.4. The following Theorem 2.3.1 states some properties of (2.42).

Theorem 2.3.1 *For any Schrödinger multiplier Φ in Corollary 2.3.2, there exist weak fundamental solutions $p(s,x;t,y)$ of (2.21) and*

$$q(s,x;t,y) = \frac{1}{\Phi(s,x)}\, p(s,x;t,y)\, \Phi(t,y)\quad on\ \ D(\Phi)^2\ \ with\ \ s < t \qquad (2.44)$$

of (2.24), i.e., p and q satisfy (2.21) and (2.24), respectively, as functions of (s,x) and their adjoint equations as functions of (t,y) in the sense of distributions.

The transformed process $(X_t,\ Q_{(s,x)},\ (s,x) \in D(\Phi),\ s \leq t \leq b)$ in (2.42) has the probability transition density q in (2.44) and it can be represented as

$$X_t = x + \int_s^t \sigma(v, X_v)\, dW_v + \int_s^t \{\beta(v, X_v) + \alpha\, \frac{\nabla^T \Phi}{\Phi}(v, X_v)\}\, dv \qquad (2.45)$$

$s \leq t \leq b$, $Q_{(s,x)}$-a.s. for λ^d-a.e. $x \in I\!\!R^d$ with $(s,x) \in D(\Phi)$ where $(W_t)_{s \leq t \leq b}$ is the d-dimensional Brownian motion w.r.t. $Q_{(s,x)}$.

Proof. Let us follow a staircase which starts from the construction of the density $p(s,x;t,y)$, passes the verification of its Chapman-Kolmogorov equation and its properties as a function of (s,x) and meets the representation of the transformed process (2.42) in terms of (2.45). It culminates with the proof of the properties of $p(s,x;t,y)$ and $q(s,x;t,y)$ as functions of (t,y).

In a first step we are going to derive $p(s,x;t,y)$ as a measurable function of $(s,x) \in D(\Phi)$ such that

$$P_{(s,x)}[f\, \Phi(t, X_t) \exp\{\int_s^t c(v, X_v)\, dv\}; t < \tau_s] \qquad (2.46)$$

$$= \int p(s,x;t,y)\, dy\, f(t,y)\, \Phi(t,y)$$

for bounded continuous functions $f(t, .\,)$ defined on $\{y \in I\!\!R^d : (t,y) \in D(\Phi)\}$. Since the left-hand side of (2.46) is a linear functional of such functions $f(t, .\,)$, the Riesz-Markov theorem yields for every $(s, x) \in D(\Phi)$ a non-negative measure $m_{s,x,t,\Phi}(dy)$ concentrated on $\{y \in \mathbb{R}^d : (t, y) \in D(\Phi)\}$ such that

$$P_{(s,x)}[f\,\Phi(t, X_t) \exp\{\int_s^t c(v, X_v)\,dv\}; t < \tau_s] \tag{2.47}$$
$$= \int f(t, y)\,m_{s,x,t,\Phi}(dy).$$

The quantity $m_{s,x,t,\Phi}(\,.\,)$ is jointly (s, x, t)-measurable on $D(\Phi) \times (s, b)$. In fact, the left-hand side of (2.47) is jointly continuous in $(s, x) \in D(\Phi)$ for every fixed $t \in (a, b)$ and right-continuous in t, $t \geq s$, for every fixed $(s, x) \in D(\Phi)$, which are consequences of Definition 2.2.2, Definition 2.3.1 and the assumptions adopted from Corollary 2.3.2. Thus Lemma 2.3.1 yields the desired measurability.

Let us refer to (3.5) where the transition density $g^o(s, x; t, y)$ is introduced. For fixed $(s, x) \in D(\Phi)$ and $t \in (s, b)$, $g^o(s, x; t, y)\,\Phi(t, y)\,dy$ can be considered as a non-negative measure concentrated on $\{y \in \mathbb{R}^d : (t, y) \in D(\Phi)\}$. Lemma 3.2.1 claims that $g^o(s, x; t, y)$ is jointly continuous in $(s, x) \in D(\Phi)$ and continuous in $t \in \{u \in (s, b) : \exists\, y \in \mathbb{R}^d \text{ such that } (u, y) \in D(\Phi)\}$ for every fixed $(s, x) \in D(\Phi)$. Hence the function $g^o(s, x; t, y)\,\Phi(t, y)$ is (s, x, t)-jointly measurable on $D(\Phi) \times (s, b)$ as a consequence of Lemma 2.3.1.

The canonical σ-algebra associated with space-time stochastic processes is treated in Doob (1984)'s part 2, chap. vii, 2. In our context we claim that for every fixed $(s, x, t) \in D(\Phi) \times (s, b)$, $m_{s,x,t,\Phi}(dy)$ is absolutely continuous w.r.t. $g^o(s, x; t, y)\,\Phi(t, y)\,dy$, both considered as non-negative measures on $(\mathbb{R}^d, \mathcal{B}(\mathbb{R}^d))$. In fact, if $A \in \mathcal{B}(\mathbb{R}^d)$ with $\Phi(t, y) = 0$ for λ^d-almost all $y \in A$, then $\{t\} \times A \subset D(\Phi)^c$ and hence $g^o(s, x; t, y) = 0$ for λ^d-almost all $y \in A$ because of (3.5). Moreover, if $A \in \mathcal{B}(\mathbb{R}^d)$ with $g^o(s, x; t, y) = 0$ for λ^d-a.a. $y \in A$, then $P_{(s,x)}[1_{\{t\} \times A}(t, X_t); t < \tau_s] = 0$ implies that $P_{(s,x)}$-a.e. trajectory which starts in (s, x) and arrives at $\{t\} \times A$ does hit $D(\Phi)^c$ on its way. Consequently, for every $A \in \mathcal{B}(\mathbb{R}^d)$ with $g^o(s, x; t, y)\,\Phi(t, y) = 0$ for λ^d-a.a. $y \in A$, (2.47) yields

$$m_{s,x,t,\Phi}(A) = P_{(s,x)}[1_{\{t\} \times A}\,\Phi(t, X_t) \exp\{\int_s^t c(v, X_v)\,dv\}; t < \tau_s] = 0.$$

The integrability assumptions adopted from Corollary 2.3.2 imply that the measures $m_{s,x,t,\Phi}(dy)$ and $g^o(s, x; t, y)\,\Phi(t, y)\,dy$ on $(\mathbb{R}^d, \mathcal{B}(\mathbb{R}^d))$ have finite mass for every $(s, x) \in D(\Phi)$ and $t \in (a, b)$, where the case of $m_{s,x,t,\Phi}(dy)$ follows by means of (2.38).

Let $T = \{(s, x, t) : (s, x) \in D_b(c), t \in (s, b)\}$, $\mathcal{T} = \mathcal{B}(D_b(c)) \otimes \mathcal{B}((a, b))$ and $\mathcal{F} = \mathcal{B}(\mathbb{R}^d)$ which is a separable σ-algebra. We can summarize that $(m_{s,x,t,\Phi}(dy))_{(s,x,t) \in T}$ and $(g^o(s, x; t, y) \Phi(t, y) dy)_{(s,x,t) \in T}$ are two $\mathcal{T}$-measurable families of non-negative measures on $(\mathbb{R}^d, \mathcal{F})$ with finite mass such that $m_{s,x,t,\Phi}(dy)$ is absolutely continuous w.r.t. $g^o(s, x; t, y) \Phi(t, y) dy$ for every fixed $(s, x, t) \in T$. Under such circumstances a version of the Radon-Nikodym theorem in Meyer (1966)'s chap vii (10) provides a non-negative $\mathcal{T} \otimes \mathcal{F}/\mathcal{B}(\mathbb{R})$-measurable density $r(s, x; t, y)$ such that

$$m_{s,x,t,\Phi}(dy) = r(s, x; t, y) \, g^o(s, x; t, y) \, \Phi(t, y) \, dy$$

for $(s, x) \in D(\Phi)$ and $t \in (s, b)$. Hence $p(s, x; t, y) = r(s, x; t, y) \, g^o(s, x; t, y)$ for $(s, x) \in D(\Phi)$ with $t \in (s, b)$ and $y \in \mathbb{R}^d$ yields (2.46) because of (2.47).

The second step shows the Chapman-Kolmogorov equation for the density $p(s, x; t, y)$ on $D(\Phi) \times D(\Phi)$, i.e.,

$$p(s, x; t, y) = \int_{A_u} p(s, x; u, z) \, dz \, p(u, z; t, y) \tag{2.48}$$

for $(s, x) \in D(\Phi)$ and λ^d-almost all $y \in A_t = \{y \in \mathbb{R}^d : (t, y) \in D(\Phi)\}$, where $a \leq s < u < t \leq b$. Because of (2.31), equation (2.46) can be rewritten in terms of $q(s, x; t, y)$ in (2.44) as

$$P_{(s,x)}[f(t, X_t) \, N_s^t] = \int q(s, x; t, y) \, dy \, f(t, y) \tag{2.49}$$

for $(s, x) \in D(\Phi)$ and functions $f(t, .)$ on $\{y \in \mathbb{R}^d : (t, y) \in D(\Phi)\}$ which are bounded and continuous. As a consequence, $q(s, x; t, y)$ is the transition density of (2.42). Because of (2.38), $P_{(s,x)}[f(t, X_t) \, N_s^t]$ is bounded and continuous in $(s, x) \in D(\Phi)$. Hence (2.49) yields

$$\int q(s, x; u, z) \, dz \, P_{(u,z)}[f(t, X_t) \, N_u^t] \tag{2.50}$$

$$= \quad P_{(s,x)}[f(t, X_t) \, N_s^u \, N_u^t] = P_{(s,x)}[f(t, X_t) \, N_s^t]$$

since N_s^t is multiplicative. When $P_{(u,z)}[f(t, X_t) \, N_u^t]$ in (2.50) is expressed in terms of (2.49), the factors $\Phi(u, z)$ cancel on A_u and we obtain

$$\int p(s, x; t, y) \, dy \, \Phi(t, y) \, f(t, y) =$$

$$= \int p(s,x;u,z)\, dz \int_{A_u} p(u,z;t,y)\, dy\, \Phi(t,y)\, f(t,y)$$

for $(s,x) \in D(\Phi)$. Thus (2.48) follows because of the sufficiently arbitrary functions $f(t,\,.\,)$.

As a third step we remark that the transition density $p(s,x;t,y)$ in (2.46) solves weakly as a function of $(s,x) \in D(\Phi)$ the diffusion equation in (2.21) which corresponds to (3.21). In fact, (2.46) appears in (3.20) in Section 3.3 as p^o with $f\,\Phi(t,\,.\,)$ in place of $f(t,\,.\,)$. Hence, under the integrability assumptions adopted from Corollary 2.3.2, Proposition 3.3.1 claims that (2.46) is the unique solution of the 'killed' Feynman-Kac type integral equation (3.9) and so a weak solution of (3.21) satisfying $\lim_{s \nearrow t} p(s,x) = f(t,x)\,\Phi(t,x)$ for $(t,x) \in D(\Phi)$.

The fourth step deals with an application of a generalized version of the *Maruyama-Grisanov drift transformation* given in Liptser-Shiryayev (1977)'s chapter 6. For fixed $(s,x) \in D(\Phi)$, we intend to find non-negative $P_{(s,x)}$-a.s. continuous solutions of the stochastic differential equation

$$
\begin{aligned}
dY_t &= Y_t\, \frac{\nabla\Phi}{\Phi}\, \sigma(t,X_t)\, dB_t, \qquad s \le t \le b \\
Y_s &= 1
\end{aligned}
\tag{2.51}
$$

which satisfy

$$\int_s^b \| \, Y_t\, \frac{\nabla\Phi}{\Phi}\, \sigma(t,X_t) \, \|^2 \; dt < \infty, \qquad P_{(s,x)}\text{-a.s.} \tag{2.52}$$

where $(B_t)_{s \le t \le b}$ denotes the d-dimensional Brownian motion w.r.t. $P_{(s,x)}$. Let $(V_s^{(n)})_{n \in I\!N}$ be a sequence of increasing stopping times defined by

$$V_s^{(n)} = \inf\{ u \in [s,b) : \int_s^u \| \, \frac{\nabla\Phi}{\Phi}\, \sigma(v,X_v) \, \|^2 \; dv \ge n \}.$$

We set $V_s = \lim_{n \to \infty} V_s^{(n)}$ and receive immediately that the path integral

$$\int_s^t \| \, \frac{\nabla\Phi}{\Phi}\, \sigma(v,X_v) \, \|^2 \; dv \tag{2.53}$$

takes the value infinity at time $t = V_s$ on $\{V_s < b\}$, $P_{(s,x)}$-a.s. Following the first part of Liptser-Shiryayev (1977)'s lemma 6.3,

$$V_s^{(1)} > s, \qquad P_{(s,x)}\text{-a.s.} \tag{2.54}$$

and

$$\lim_{n \to \infty} \int_s^{V_s^{(n)}} \| \, \frac{\nabla\Phi}{\Phi}\, \sigma(v,X_v) \, \|^2 \; dv = \infty, \qquad P_{(s,x)}\text{-a.s.} \tag{2.55}$$

on $\{V_s < b\}$ are necessary and sufficient for (2.51) to have a non-negative $P_{(s,x)}$-a.s. continuous solution. The condition (2.55) expresses that on $\{V_s < b\}$, 'the departure' of the integral (2.53) for infinity occurs in a continuous manner as $t \to V_s(X)$ for $P_{(s,x)}$-a.a. $X \in C([a,b], \mathbb{R}^d)$.

First we verify condition (2.54). Since $D(\Phi)$ is assumed to be open, there exists a sequence $(U_n)_{n \in \mathbb{N}}$ of open balls with common center (s,x) such that $U_n \subset \overline{U_n} \subset U_{n-1} \subset D(\Phi)$ where $\overline{U_n}$ is the compact closure of U_n. The associated sequence of decreasing stopping times

$$V_s(U_n) = \inf\{u \in [s,b) : (u, X_u) \notin U_n\}, \quad n \in \mathbb{N}$$

satisfies $s < V_s(U_{n+1}) < V_s(U_n) < \tau_s$ as well as

$$\int_s^{V_s(U_n)} \| \frac{\nabla \Phi}{\Phi} \sigma(v, X_v) \|^2 \, dv \leq \frac{\sup_D \| \sigma \|^2}{\min_{\overline{U_n}} \Phi} \int_s^{V_s(U_n)} \| \nabla \Phi(v, X_v) \|^2 \, dv < \infty$$

$P_{(s,x)}$-a.s. if n is sufficiently large as a consequence of Definition 2.2.2 and Lemma 4.3.1. Hence Beppo-Levi's theorem yields

$$\int_s^{V_s(U_n)} \| \frac{\nabla \Phi}{\Phi} \sigma(v, X_v) \|^2 \, dv \searrow 0 \quad \text{as } n \nearrow \infty, \quad P_{(s,x)}\text{-a.s.}$$

Thus there exists a number $n(X) \in \mathbb{N}$ such that $V_s(U_n)(X) \leq V_s^{(1)}(X)$ for $P_{(s,x)}$-a.a. $X \in C([a,b], \mathbb{R}^d)$. Since the continuity of paths implies $s < V_s(U_n)$, $P_{(s,x)}$-a.s. for $n \in \mathbb{N}$, (2.54) is satisfied.

Second, condition (2.55) has to be checked. Lemma 4.3.1 implies that the path integral (2.53) is $P_{(s,x)}$-a.s. finite and right-continuous for $t \in [s, \tau_s)$. Hence

$$\int_s^{V_s^{(n)}} \| \frac{\nabla \Phi}{\Phi} \sigma(v, X_v) \|^2 \, dv = n \tag{2.56}$$

follows on $\{V_s < b\}$, $P_{(s,x)}$-a.s. for $n \in \mathbb{N}$. Under the present assumptions, Corollary 2.3.2 yields

$$P_{(s,x)}[N_s^t] = 1 \quad \text{for } s \leq t \leq b.$$

Consequently, $(N_s^t)_{s \leq t < b}$ in Definition 2.3.2 is even a martingale and hence $P_{(s,x)}$-a.s. continuous according to Clark (1970). In this case Corollary 2.3.1 claims that the path integral (2.53) takes the value infinity at time $t = \tau_s$,

$P_{(s,x)}$-a.s. on $\{\tau_s < b\}$. Since τ_s is the earliest time t which realizes an infinite integral value in (2.53),

$$\tau_s = V_s, \quad P_{(s,x)}\text{-a.s. on } \{V_s < b\}.$$

Hence condition (2.55) is a consequence of (2.56).

Following the second part of Liptser-Shiryayev (1977)'s lemma 6.3, the unique non-negative $P_{(s,x)}$-a.s. continuous solution of (2.51) satisfying (2.52) is given as

$$Y_t = \exp\{\Gamma_s^t(\frac{\nabla\Phi}{\Phi}\sigma(.,X.)) - \frac{1}{2}\int_s^t \|\frac{\nabla\Phi}{\Phi}\sigma(v,X_v)\|^2 \, dv\}$$

for $s \leq t \leq b$ which is just the representation (2.35) of N_s^t in Corollary 2.3.1. Under $P_{(s,x)}[N_s^b] = 1$ provided by Corollary 2.3.2, Liptser-Shiryayev (1977)'s theorem 6.2 claims that

$$W_t = B_t - \int_s^t \sigma \frac{\nabla^T\Phi}{\Phi}(v,X_v) \, dv, \quad s \leq t \leq b \tag{2.57}$$

is a Brownian motion with respect to $Q_{(s,x)}$ defined as $dQ_{(s,x)} = N_s^b \, dP_{(s,x)}$ in (2.43) of Corollary 2.3.2. As shown in Stroock-Varadhan (1969)'s section 3, the d-dimensional Brownian motion $(B_t)_{s\leq t\leq b}$ w.r.t. $P_{(s,x)}$ in Definition 2.3.1 can be expressed as

$$B_t = \int_s^t \sigma^{-1}(v,X_v) \, dX_v - \int_s^t \sigma^{-1}\beta(v,X_v) \, dv \tag{2.58}$$

for $s \leq t \leq b$, $P_{(s,x)}$-a.s., where σ^{-1} is the inverse of σ, the positive definite symmetric square root of the diffusion matrix α. Combining (2.57) and (2.58), we receive

$$W_t = \int_s^t \sigma^{-1}(v,X_v) \, dX_v - \int_s^t \{\sigma^{-1}\beta(v;X_v) + \sigma\frac{\nabla^T\Phi}{\Phi}(v,X_v)\} \, dv$$

for $s \leq t \leq b$, $Q_{(s,x)}$-a.s., which yields representation (2.45).

In the fifth and last step of this proof it remains to show that $p(s,x;t,y)$ constructed in the first step and $q(s,x;t,y)$ in (2.44) are weak fundamental solutions of (2.21) and (2.24), respectively. The properties of $p(s,x;t,y)$ as a function of (s,x) were proved in the third step. They yield the corresponding properties of $q(s,x;t,y)$ as a function of (s,x) by means of Lemma 2.2.2. The way of conclusion is now going to be reversed; we first show the properties of $q(s, x; t, y)$ as a function of (t, y) and then use them to deduce the properties of $p(s, x; t, y)$ as a function of (t, y).

Let us apply the classical Itô formula to $\gamma \in C^{1,2}_{comp}([s, b) \times I\!R^d)$ generated along the space-time semi-martingale ($(t, X_t)_{s \leq t \leq b}$, $Q_{(s,x)}$) in (2.42). By means of Lemma 4.3.1 and Schwarz's inequality,

$$\int_s^t \| \beta(v, X_v) + \alpha \, \frac{\nabla^T \Phi}{\Phi}(v, X_v) \| \, dv$$

is verified to be a function of $t \in [s, b]$ of finite variation, $Q_{(s,x)}$-a.s. Hence

$$\gamma(t, X_t) - \gamma(s, x) = \int_s^t \nabla \gamma \, \sigma(v, X_v) \, dW_v$$

$$+ \int_s^t (\frac{\partial}{\partial v} \gamma(v, X_v) + \frac{1}{2} \sum_{i,j=1}^d \alpha_{ij} \frac{\partial^2}{\partial x_i \, \partial x_j} \gamma(v, X_v)$$

$$+ \nabla \gamma(v, X_v) \cdot \{\beta(v, X_v) + \alpha \, \frac{\nabla^T \Phi}{\Phi}(v, X_v)\}) \, dv$$

where

$$(\int_s^t \nabla \gamma \, \sigma(v, X_v) \, dW_v)_{s \leq t \leq b}$$

is a square-integrable, $Q_{(s,x)}$-a.s. continuous martingale. Taking the expectation expressed in terms of $q(s, x; t, y)$ in (2.44) yields

$$\int q(s, x; t, y) \, dy \, \gamma(t, y) - \gamma(s, x) \qquad (2.59)$$

$$= \int_s^t dv \int q(s, x; v, y) \, dy$$

$$\cdot (\frac{\partial}{\partial v} \gamma(v, y) + \frac{1}{2} \sum_{i,j=1}^d \alpha_{ij}(v, y) \frac{\partial^2}{\partial x_i \, \partial x_j} \gamma(v, y)$$

$$+ \nabla \gamma(v, y) \cdot \{\beta(v, y) + \alpha(v, y) \, \frac{\nabla^T \Phi}{\Phi}(v, y)\})$$

for $s \leq t \leq b$. On the other hand, we can verify analogously to the proof of Lemma 2.2.2 that

$$\int dy \, \nabla \gamma(v, y) \, \alpha(v, y) \frac{\nabla^T \Phi}{\Phi}(v, y) \, \kappa(v, y) \qquad (2.60)$$

$$= \int dy \, \gamma(v, y) \cdot$$

$$(\frac{-\kappa}{\Phi}(v,y)\,\triangle_\alpha(v,y)\,\Phi(v,y) + \kappa(v,y)\,\frac{\nabla\Phi}{\Phi}(v,y)\,\alpha(v,y)\,\frac{\nabla^T\Phi}{\Phi}(v,y)$$

$$+ \frac{-\kappa}{\Phi}(v,y)\sum_{i,j=1}^{d}(\frac{\partial}{\partial x_j}\alpha_{ij}(v,y))\,D_i\Phi(v,y) - \frac{\nabla\Phi}{\Phi}(v,y)\,\alpha(v,y)\,\nabla^T\kappa(v,y))$$

for $s \le v \le b$, where $\kappa \in C^{1,2}([s,b)\times I\!\!R^d)$ plays the role of a function $q(s,x;t,y)$ which is differentiable in (t,y). In fact, we may assume $\operatorname{supp}\gamma \subset D(\Phi)$ since the trajectories $(t,X_t)_{s\le t\le b}$ live $Q_{(s,x)}$-a.s. in the set $D(\Phi)$ as shown in Corollary 2.3.2. Hence Φ can be approximated in the $H^{1,2}_{loc}$-sense as well as uniformly on $\operatorname{supp}\gamma$ by a sequence $(\Phi_n)_{n\in I\!\!N} \subset C^{1,2}(\operatorname{supp}\gamma)$. An addition of

$$\int_s^b dv \int dy\, L\,\gamma(v,y)\,\kappa(v,y)$$

$$= -\int dy\,\gamma(s,y)\,\kappa(s,y) + \int_s^b dv \int dy\,\gamma(v,y)\,L^*\kappa(v,y)$$

and (2.60) yields

$$\int dy\,\gamma(s,y)\,\kappa(s,y) \tag{2.61}$$

$$+ \int_s^b dv \int dy\,(\frac{\partial}{\partial v} + \frac{1}{2}\sum_{i,j=1}^{d}\alpha_{ij}(v,y)\,\frac{\partial^2}{\partial x_i\,\partial x_j}$$

$$+ \{\beta(v,y) + \alpha\,\frac{\nabla^T\Phi}{\Phi}(v,y)\}\cdot\nabla)\,\gamma(v,y)\,\kappa(v,y)$$

$$= \int_s^b dv \int dy\,\gamma(v,y)\,(L^* - \frac{1}{\Phi}\operatorname{tr}(\alpha\,J\Phi)(v,y) + \frac{\nabla\Phi}{\Phi}\,\alpha\,\frac{\nabla^T\Phi}{\Phi}(v,y)$$

$$- \frac{1}{\Phi}\sum_{i,j=1}^{d}(\frac{\partial}{\partial x_j}\alpha_{ij}(v,y))\,D_i\Phi(v,y) - \frac{\nabla\Phi}{\Phi}\,\alpha(v,y)\,\nabla^T)\kappa(v,y).$$

Finally we conclude from (2.59) that the left-hand side of (2.61) becomes zero when we replace $\kappa(v,y)$ by $q(s,x;v,y)$ for $(v,y) \in [s,b)\times I\!\!R^d$. In order to prove that $p(s,x;t,y)$ as a function of (t,y) is a weak solution of

$$L^*p + c\,p = 0 \quad \text{on } D(\Phi)$$

we express it in terms of $q(s,x;t,y)$ in (2.44) and apply the just shown properties of $q(s,x;t,y)$ as a function of (t,y). The required approximation procedure can be managed similarly as in Lemma 2.2.2. $\diamond$

Lemma 2.3.1 *Let $T = (a, b)$, $\mathcal{T}$ its Borel's σ-algebra and $(\Omega, \mathcal{F})$ a measurable space. The product space $T \times \Omega$ and $\mathbb{R}$, respectively, are endowed with the canonical product σ-algebra*

$$\mathcal{T} \otimes \mathcal{F} = \sigma(A \in \mathcal{T} \times \mathcal{F})$$

and Borel's σ-algebra $\mathcal{B}(\mathbb{R})$. If the function

$$f : \left\{ \begin{array}{ccc} T \times \Omega & \to & \mathbb{R} \\ (t, \omega) & \longmapsto & f(t, \omega) \end{array} \right.$$

satisfies

$$\begin{array}{ccl} t & \to & f(t, \omega) \quad \text{is right-continuous for every } \omega \in \Omega \\ \omega & \to & f(t, \omega) \quad \text{is } \mathcal{F}/\mathcal{B}(\mathbb{R})\text{-measurable for every } t \in (a, b) \end{array}$$

then it is jointly measurable in $(t, \omega) \in T \times \Omega$, i.e., f is $\mathcal{T} \otimes \mathcal{F}/\mathcal{B}(\mathbb{R})$-measurable.

Proof. For partitions $a < t_0 < t_1 < \ldots < t_n < b$ with the maximal size of increments denoted by $\Delta = \max\{(| t_i - t_{i-1} |)_{1 \leq i \leq n}, | t_0 - a |, | b - t_n |\}$, we define step functions

$$f_n(t, \omega) = \sum_{i=1}^{n} f(t_{i-1}, \omega) \, 1_{[t_{i-1}, t_i)}(t) + f(t_n, \omega) \, 1_{[t_n, b)}(t)$$

which are $\mathcal{T} \otimes \mathcal{F}/\mathcal{B}(\mathbb{R})$-measurable. The right-continuity of f in the variable t provides $\lim_{\Delta \to 0} f_n(t, \omega) = f(t, \omega)$ on $T \times \Omega$. Hence f is $\mathcal{T} \otimes \mathcal{F}/\mathcal{B}(\mathbb{R})$-measurable as the limit of $\mathcal{T} \otimes \mathcal{F}/\mathcal{B}(\mathbb{R})$-measurable functions f_n. $\Diamond$

2.4   Schrödinger processes

In Section 1.5 we introduced the Schrödinger-Nagasawa representation of diffusions based on a time-symmetrical view of diffusions proposed in Schrödinger (1931) and investigated in Nagasawa (1989). Diffusions given in this time-symmetrical representation were briefly called Schrödinger processes. Now we formally state

Definition 2.4.1 The *Schrödinger process* for a pair of non-negative functions $\hat{\varphi}$ and φ on $[a, b] \times \mathbb{R}^d$, $-\infty < a < b < \infty$, and a transition density $p(s, x; t, y)$, $a \leq s < t \leq b$, $x, y \in \mathbb{R}^d$, is a space-time diffusion process

$$(\, (t, X_t) \, , \, a \leq t \leq b, \, Q \,)$$

determined by its finite dimensional distribution densities

$$\hat{\varphi}(a,x)\,dx\,p(a,x;t_1,y_1)\,dy_1 \cdot \ldots \cdot p(t_n,y_n;b,z)\,dy_n\,\varphi(b,z)\,dz \tag{2.62}$$

for $a < t_1 < \ldots < t_n < b$, $n \in I\!N$, where

$$\varphi(t,y) = \int_{I\!R^d} p(t,y;b,z)\,dz\,\varphi(b,z) \quad \text{for } a \leq t < b \tag{2.63}$$

$$\hat{\varphi}(t,y) = \int_{I\!R^d} \hat{\varphi}(a,x)\,dx\,p(a,x;t,y) \quad \text{for } a < t \leq b \tag{2.64}$$

and

$$\int_{I\!R^d} \int_{I\!R^d} \hat{\varphi}(a,x)\,dx\,p(a,x;b,z)\,dz\,\varphi(b,z) = 1. \tag{2.65}$$

We notice that $\hat{\varphi}\,\varphi(t, \, . \,)$ is the distribution density of a Schrödinger process at time $t \in [a,b]$ where φ and $\hat{\varphi}$ are space-time harmonic and co-harmonic functions of $p(s,x;t,y)$, respectively. As a transition density, $p(s,x;t,y)$ satisfies the Chapman-Kolmogorov equation, but it is not assumed to be normalized. Hence a Schrödinger process Q is a diffusion process with creation and killing which is however mass preserving, i.e., conservative, because of (2.65). In order to determine a Schrödinger process Q, it is sufficient to prescribe non-negative functions $\varphi(b, \, . \,)$ and $\hat{\varphi}(a, \, . \,)$ as well as a transition density $p(s,x;t,y)$ such that (2.65) is satisfied. Then, φ and $\hat{\varphi}$ can be defined on $[a,b] \times I\!R^d$ according to (2.63) and (2.64), respectively.

Theorem 2.4.1 is going to provide the construction of a Schrödinger process Q in terms of a Schrödinger multiplier Φ in Definition 2.2.2, the parabolic differential operator L in (2.1) and a non-negative measurable function $\hat{\varphi}(a, \, . \,)$ satisfying (2.66). Actually, Theorem 2.3.1 yields a transition density $p(s,x;t,y)$ which is a weak fundamental solution of (2.21) with potential c in (2.22) where (2.63) is satisfied because of (2.31), (2.41) and (2.46). The initial distribution density of Q can be chosen freely on $\{\Phi(a, \, . \,) > 0\}$ as $\hat{\varphi}(a, \, . \,)\,\Phi(a, \, . \,)$ by means of a non-negative $\hat{\varphi}(a, \, . \,)$ such that

$$\int_{I\!R^d} \hat{\varphi}(a,x)\,\Phi(a,x)\,dx = 1. \tag{2.66}$$

Extended according to (2.64), $\hat{\varphi}$ is a weak solution of the adjoint equation of (2.21). We notice that the construction of a diffusion process with singular drift in Corollary 2.3.2 is a more detailed result than its application to Schrödinger processes. In fact, the transformation by means of $(N_s^t)_{a \leq s \leq t \leq b}$

shows us in Theorem 2.3.1 advantageously much of the pathwise behavior of
the process $Q_{(s,x)}$ in (2.42) starting from space-time points $(s, x) \in D(\Phi)$.

Theorem 2.4.1 *Let Φ be a Schrödinger multiplier in Definition 2.2.2, c its reference potential in (2.22) and $\hat{\varphi}(a, .)$ a non-negative measurable function which
satisfies the assumptions of Corollary 2.3.2 and (2.66). Then the space-time
diffusion process*

$$((t, X_t), \ a \le t \le b, \ Q) \tag{2.67}$$

defined by means of $dQ_{(a,x)} = N_a^b \, dP_{(a,x)}$ in (2.42) as

$$Q(A) = \int_{I\!\!R^d} \hat{\varphi}(a, x) \, \Phi(a, x) \, dx \, Q_{(a,x)}[1_A] \tag{2.68}$$

for $A \in \sigma((v, X_v), \ a \le v \le b)$ is a Schrödinger process *in Definition 2.4.1
determined by*

$$\{ \, \hat{\varphi}(a, .), \ \Phi, \ L \, \}$$

*where the differential operator L is given in (2.1) and $\hat{\varphi}$ is extended to $[a, b] \times I\!\!R^d$
according to (2.64).*

The Schrödinger process (2.67) has the transition density

$$q(s, x; t, y) = \frac{1}{\Phi(s, x)} \, p(s, x; t, y) \, \Phi(t, y) \quad on \ \ D(\Phi)^2 \ \ with \ \ s < t \tag{2.69}$$

*in (2.44) where $D(\Phi)$ is given in Definition 2.2.2. Moreover, (2.67) is a dif-
fusion process with diffusion matrix α in the Laplacian of α-divergence type
in (2.2) and with in general singular drift*

$$\frac{1}{2} \alpha_\nabla + \beta + \alpha \, \frac{\nabla^T \Phi}{\Phi} \tag{2.70}$$

*where α_∇ is in (2.3). The space-time trajectories $(t, X_t)_{a \le t \le b}$ of (2.67) do not
hit the set of zeros and singularities*

$$(\, D(\Phi) \cap D(\hat{\Phi}) \,)^c$$

*Q-a.s., provided $\hat{\varphi}$ is a Schrödinger multiplier of Definition 2.2.2 and may
therefore be denoted by $\hat{\Phi}$.*

Let the time-reversal transformation *be defined for space-time functions f
on $[a, b] \times I\!\!R^d$ as*

$$f^{\#}(t, y) = f(a + b - t, y). \tag{2.71}$$

Then the time-reversed Schrödinger process

$$((t, \hat{X}_t = X_{a+b-t}), \ a \le t \le b, \ Q) \tag{2.72}$$

evolving with increasing time t is again a Schrödinger process, now determined by

$$\{ \ \Phi^\# , \ \hat{\Phi}^\# , \ p^\#(s, x; t, y) \ \}$$

where $p^\#(s, x; t, y) = p(a + b - t, y; a + b - s, x)$. *It possesses the distribution density*

$$\hat{\rho}(t, y) \ = \ \Phi^\#(t, y) \, \hat{\Phi}^\#(t, y) \quad \text{for } t \in [a, b] \tag{2.73}$$

and the transition density

$$\hat{q}(s, x; t, y) = \frac{1}{\hat{\Phi}^\#(s, x)} \, p^\#(s, x; t, y) \, \hat{\Phi}^\#(t, y) \quad \text{on } (D(\hat{\Phi})^\#)^2 \tag{2.74}$$

with $s < t$, *where* $\hat{\Phi}^\#$ *and* $\Phi^\#$ *are space-time harmonic and co-harmonic functions of* $p^\#(s, x; t, y)$, *respectively. Moreover, the time-reversed Schrödinger process (2.72) is a diffusion process* $((t, X_t), \ a \le t \le b, \ \hat{Q})$ *described by*

$$\hat{Q}(A) = \int_{\mathbb{R}^d} \hat{\Phi}(a, x) \, \Phi(a, x) \, dx \, P_{(a,x)}[\hat{N}_a^b \, 1_A] \tag{2.75}$$

for $A \in \sigma((v, X_v), \ a \le v \le b)$ *in terms of*

$$\hat{N}_s^t(X) = \frac{\hat{\Phi}^\#(t, X_t)}{\hat{\Phi}^\#(s, X_s)} \, \exp\{ \int_s^t (-\frac{L^*\hat{\Phi}}{\hat{\Phi}})^\#(v, X_v) \, dv \} \, 1_{\{t < \tau_s\}} \tag{2.76}$$

for $a \le s < t \le b$, *which is the dual multiplicative functional to* N_s^t *in Definition 2.3.2. As a consequence, (2.72) possesses the diffusion matrix* $\alpha^\#$ *and the in general singular drift*

$$\frac{1}{2} \alpha_\nabla^\# - \beta^\# + \alpha^\# \frac{\nabla^T \hat{\Phi}^\#}{\hat{\Phi}^\#}. \tag{2.77}$$

Remark 2.4.1 The properties of Schrödinger processes in Definition 2.4.1 provided by Theorem 2.4.1 give rise to interpretations as given in Section 1.3. Recalling Schrödinger (1931)'s original investigations, we notice that the distribution density of a Schrödinger process Q at time $t \in [a, b]$ is still the product of the Schrödinger multipliers $\hat{\Phi}$ and Φ at time t. However, dynamics is now given by $L + c \, I$ where the reference potential c in (2.22) is in general unbounded and even not continuous. The class of Schrödinger processes is invariant under time-reversal. In fact, $\Phi^{\#}$ and $\hat{\Phi}^{\#}$ play under time-reversal the rôles of $\hat{\Phi}$ and Φ, respectively. The justification of the notation ' $\#$ ' (sharp) beside ' $\char`\^$ ' (hat) follows from $\hat{q} \neq q^{\#}$ which will be evident after the calculation of $\hat{q}$ in (2.78).

Remark 2.4.2 The roles of the Schrödinger multiplier Φ and its dual $\hat{\Phi}$ in the construction of Schrödinger processes in Theorem 2.4.1 can be interchanged. In fact, $L\Phi/\Phi = L^*\hat{\Phi}/\hat{\Phi}$ holds because of (2.64). Moreover, relation (2.46) and the integrability conditions in Corollary 2.3.2 can be given in terms of $\hat{\Phi}$, employing a formulation in terms of time-reversal.

Remark 2.4.3 In case of Theorem 2.4.1, Corollary 2.3.2 yields $Q(b < \tau_a) = 1$. This is in accordance with the continuous version of the Schrödinger system (1.17) & (1.18) which represents continuity conditions having their origin in (1.2) and (1.3), respectively. They express that paths leaving at time a must arrive at time b under the measure Q. In particular, such paths must not cross the zero set $\{\hat{\Phi}\, \Phi = 0\}$ since the drift vectors in (2.70) and (2.77) are only well-defined on $D(\hat{\Phi}) \cap D(\Phi)$.

Remark 2.4.4 Schrödinger processes related to Schrödinger equations (2.5) are provided by Theorem 2.4.1 with Φ given in (2.9). Solutions Ψ of (2.5) have to satisfy general conditions found in Simon (1982) and Carmona (1985) which were discussed in the context of Theorem 2.1.1. The properties of $p(s, x; t, y)$ in Theorem 2.3.1 imply that $\hat{\Phi}$ in (2.12) is just the function defined by (2.64) for the given $\hat{\Phi}(a, \, . \,)$.

Proof. Let us verify Definition 2.4.1 in case of the space-time stochastic process (2.67) defined by (2.68). Given Φ, the transition density $p(s, x; t, y)$ of Theorem 2.3.1 is a weak fundamental solution of the diffusion equation (2.21) with singular potential c in (2.22). Hence (2.63) follows from (2.31), (2.41), (2.43) and (2.44). Moreover, (2.64) holds by the definition of $\hat{\varphi}$ and (2.65) is a consequence of assumption (2.66) and (2.63) for $t = a$. Because of (2.68), the process (2.67) has the transition probability (2.42) with transition density (2.44), i.e., (2.69). Hence the finite dimensional distributions of (2.67) are of the form (2.62).

The representation of (2.67) as a diffusion process with diffusion matrix α and singular drift (2.70) is a consequence of Theorem 2.3.1, particularly of (2.45). In fact, for λ-a.a. $x \in I\!\!R^d$ with $(a, x) \in D(\Phi)$, the process (2.42) has $Q_{(a,x)}$-a.s. continuous space-time trajectories which do not hit $D(\Phi)$, $Q_{(a,x)}$-a.s. The singularities of $\hat{\Phi}$ defined in (2.64) are contained in $D(\Phi)$ since $\hat{\Phi}$ inherits them, but generally not its zeros, from Φ through $p(s, x; t, y)$. We recall from Definition 2.2.2 that $(D(\hat{\Phi}) \cap D(\Phi))^c$ is closed and from Definition 2.4.1 that $\hat{\Phi}\,\Phi(t, .)$ is the distribution density of (2.67). As a consequence, the continuous space-time paths of (2.67) do not hit $(D(\hat{\Phi}) \cap D(\Phi))^c$, Q-a.s.

Let us deduce the properties of the time-reversal (2.72) of (2.67). The distribution density $\hat{\rho}$ in (2.73) is an obvious consequence of (2.71) and (2.72). The transition density $\hat{q}(s, x; t, y)$ in (2.74) is provided by Williams (1979)'s vol. 1 chap. III.69 ('Nagasawa's formula') as

$$\hat{q}(s, x; t, y) \hspace{4cm} (2.78)$$

$$= \hat{\Phi}\,\Phi(a + b - t, y)\, q(a + b - t, y; a + b - s, x)\, \frac{1}{\hat{\Phi}\,\Phi(a + b - s, x)}$$

$$= \hat{\Phi}(a + b - t, y)\, p(a + b - t, y; a + b - s, x)\, \frac{1}{\hat{\Phi}(a + b - s, x)}, \quad s < t$$

for (s, x), $(t, y) \in D(\hat{\Phi})^\#$ where $D(\hat{\Phi})^\# = \{(v, y) : (a + b - v, y) \in \{\rho > 0\}\}$. We notice that $\hat{q} \neq q^\#$ which justifies our notation as mentioned in Remark 2.4.1. It is straightforward to verify that (2.72) is again a Schrödinger process of Definition 2.4.1 where $\Phi^\#$, $\hat{\Phi}^\#$ and $p^\#(s, x; t, y)$ play the roles of $\hat{\Phi}$, Φ and $p(s, x; t, y)$, respectively.

In the final part of this proof we show a probabilistical and an analytical way how to identify the drift (2.77) of (2.72). When starting with a probabilistical argument, we notice first that a diffusion process $P_{(a,x)}$ with a diffusion matrix 1 and drift vector $\beta = 0$ may be assumed. In fact, the case of a general elliptic diffusion matrix α follows from a change of time scale which can be found in the proof of Theorem 4.3.1 as an application of Kunita (1990)'s § 3.2. Moreover, a drift β which satisfies the Novikov condition (1.51) can always be added by means of the Maruyama-Girsanov drift transformation. Hence let us consider the family $W_{(s,x)}$, $(s, x) \in [a, b] \times I\!\!R^d$, of probability measures in Definition 2.3.1 associated with L in (2.1) for $\alpha = 1$ and $\beta = 0$ which is just the Wiener measure on $C([a, b], I\!\!R^d)$. The transition density $\hat{q}(s, x; t, y)$ in (2.74) of (2.72) yields

$$Q[f(\hat{X}_s, \hat{X}_t)] = \int \int \hat{\Phi}^\#(s, x)\, dx\, \Phi^\#(s, x)\, \hat{q}(s, x; t, y)\, dy\, f(x, y) =$$

$$= \int\int \hat{\Phi}(a+b-s,x)\,\Phi(a+b-s,x)\,dx\,(1/\hat{\Phi}(a+b-s,x))$$

$$\cdot p(a+b-t,y;a+b-s,x)\,\hat{\Phi}(a+b-t,y)\,f(x,y)\,dy$$

$$= \int\int \hat{\Phi}(\tilde{s},y)\,\Phi(\tilde{t},x)\,dx$$

$$\cdot W_{(\tilde{s},y)}[f(X_{\tilde{t}},X_{\tilde{s}})\,\exp\{\int_{\tilde{s}}^{\tilde{t}} c(v,X_v)\,dv\}\,1_{\{\tilde{t}<\tau_{\tilde{s}}\}}\,1_{\{X_{\tilde{t}}\in dx\}}]\,dy$$

for $\tilde{s} = a+b-t$, $\tilde{t} = a+b-s$ and bounded measurable functions $f(x,y)$ on $\mathbb{R}^d \times \mathbb{R}^d$ where (2.46) has been employed. Since

$$W_{(s,x)}[g(v,X_v)\,1_{\{X_t\in dy\}}] = W_{(a+b-t,y)}[g(v,X_{a+b-v})\,1_{\{X_{a+b-s}\in dx\}}]$$

for bounded space-time measurable functions g, we can write

$$Q[f(\hat{X}_s,\hat{X}_t)] = \int\int \hat{\Phi}(\tilde{s},y)\,\Phi(\tilde{t},x)\,dx\,W_{(a+b-\tilde{t},x)}[f(X_{a+b-\tilde{t}},X_{a+b-\tilde{s}})$$

$$\cdot \exp\{\int_{\tilde{s}}^{\tilde{t}} c(v,X_{a+b-v})\,dv\}\,1_{\{a+b-\tilde{s}<\tau_{a+b-\tilde{t}}\}}\,1_{\{X_{a+b-\tilde{s}}\in dy\}}]\,dy$$

$$= \int\int \hat{\Phi}^{\#}(s,x)\,\Phi^{\#}(s,x)\,dx\,W_{(s,x)}[f(X_s,X_t)\,\frac{\hat{\Phi}^{\#}(t,X_t)}{\hat{\Phi}^{\#}(s,X_s)}$$

$$\cdot \exp\{\int_s^t c^{\#}(v,X_v)\,dv\}\,1_{\{t<T_s\}}\,1_{\{X_t\in dy\}}]\,dy.$$

Thus (2.76) follows because of $L^*\hat{\Phi}/\hat{\Phi} = -c$ which is a consequence of (2.64). Finally, (2.77) is obtained by an application of Theorem 2.3.1.

The analytical way is to show that $\hat{q}(s,x;t,y)$ in (2.74) as a function of $(s,x) \in (D(\hat{\Phi})\cap D(\Phi))^{\#}$ satisfies (2.81) which has the drift coefficient (2.77). Following Theorem 2.3.1, $q(s,x;t,y)$ in (2.69) as a function of (t,y) and the distribution density

$$\rho(t,y) = \int_{\mathbb{R}^d} \hat{\Phi}(a,x)\,dx\,p(s,x;t,y)\,\Phi(t,y)$$

of (2.67) are weak solutions of

$$L^*q - \frac{q}{\Phi}\,\triangle_\alpha\,\Phi + q\,\frac{\nabla\Phi}{\Phi}\,\alpha\,\frac{\nabla\Phi^T}{\Phi} + \frac{q}{\Phi}\sum_{i,j=1}^{d}(\frac{\partial}{\partial x_j}a_{ij})\,\Phi_i - \frac{\nabla\Phi}{\Phi}\,\alpha\,\nabla^T q = 0. \quad (2.79)$$

We write $\iota = a+b$ and notice that

$$\hat{q}(s,x;t,y) = \hat{\Phi}\,\Phi(\iota-t,y)\,\hat{r}(s,x)$$

for

$$\hat{r}(s,x) = \frac{q(\iota - t, y; \iota - s, x)}{\hat{\Phi}\, \Phi(\iota - s, x)} \tag{2.80}$$

because of (2.78) and (2.80), where $s < t$ and $(s,x), (t,y) \in (D(\hat{\Phi}) \cap D(\Phi))^{\#}$. Consequently, the fraction (2.80) is claimed to be a weak solution of

$$-\frac{\partial}{\partial s}p + \frac{1}{2}\nabla(\alpha^{\#}\nabla p) - \beta^{\#}\cdot\nabla p + \frac{\hat{\Phi}^{\#}_{\nabla}}{\hat{\Phi}^{\#}}\alpha^{\#}\nabla^{T}p = 0. \tag{2.81}$$

The verification consists of an approximation procedure where we replace $\hat{\Phi}$ in (2.81) by ρ/Φ. Let $\gamma \in C^{\infty}_{comp}((D(\hat{\Phi}) \cap D(\Phi))^{\#})$ be fixed. Due to Definition 2.2.2, there exist sequences $(\hat{\Phi}_n)_{n\in\mathbb{N}}$ and $(\Phi_n)_{n\in\mathbb{N}}$ in $C^{1,2}((\operatorname{supp}\gamma)^{\#})$ which approximate $\hat{\Phi}$ and Φ, respectively, uniformly as well as in the $H^{1,2}_{loc}$-sense on $(\operatorname{supp}\gamma)^{\#}$. Since $q(v,z;\cdot,.) \in L^1_{loc}(D(\hat{\Phi}) \cap D(\Phi))$ for every fixed $(v,z) \in D(\hat{\Phi}) \cap D(\Phi)$, it can be approximated in the $L^1_{loc}(D(\hat{\Phi}) \cap D(\Phi))$-sense by a sequence $(q_m)_{m\in\mathbb{N}}$ in $C^{1,2}(D(\hat{\Phi}) \cap D(\Phi))$. As an immediate consequence, $(\hat{r}^m_n)_{(m,n)\in\mathbb{N}^2} \subset C^{1,2}((D(\hat{\Phi}) \cap D(\Phi))^{\#})$ determined by

$$\hat{r}^m_n(s,x) = \frac{q_m(\iota - s, x)}{\hat{\Phi}_n\, \Phi_n(\iota - s, x)} \tag{2.82}$$

for $(s,x) \in (D(\hat{\Phi}) \cap D(\Phi))^{\#}$ approximates $\hat{r} \in L^1(\operatorname{supp}\gamma)$ in (2.80) in the $L^1(\operatorname{supp}\gamma)$-sense. Simultaneously, $(\rho_n)_{n\in\mathbb{N}}$ determined by $\rho_n = \hat{\Phi}_n\, \Phi_n$ approximates the distribution density $\rho = \hat{\Phi}\, \Phi$ uniformly on $(\operatorname{supp}\gamma)^{\#}$. In order to show that $\hat{r}$ in (2.80) is a weak solution of (2.81), integrations by parts need to be applied to

$$\int ds \int dx\, \gamma(s,x)\, [\,-\frac{\partial}{\partial s}(\frac{q_m}{\rho_n})(\iota - s, x)$$

$$+\frac{1}{2}\sum_{i,j=1}^{d} a_{ij}(\iota - s, x)\, \frac{\partial^2}{\partial x_i \partial x_j}\frac{q_m(\iota - s, x)}{\rho_n(\iota - s, x)}$$

$$+\frac{1}{2}\sum_{i,j=1}^{d} (\frac{\partial}{\partial x_j}a_{ij}(\iota - s, x))\, \frac{\partial}{\partial x_i}\frac{q_m(\iota - s, x)}{\rho_n(\iota - s, x)}$$

$$-\sum_{i=1}^{d} \beta_i(\iota - s, x)\, \frac{\partial}{\partial x_i}\frac{q_m(\iota - s, x)}{\rho_n(\iota - s, x)}$$

$$-\frac{1}{\Phi_n(\iota - s, x)}\sum_{i,j=1}^{d} a_{ij}(\iota - s, x)\, \frac{\partial}{\partial x_i}\Phi_n(\iota - s, x)\, \frac{\partial}{\partial x_j}\frac{q_m(\iota - s, x)}{\rho_n(\iota - s, x)}$$

$$+ \frac{1}{\rho_n(\iota - s, x)} \sum_{i,j=1}^{d} a_{ij}(\iota - s, x) \frac{\partial}{\partial x_i} \rho_n(\iota - s, x) \frac{\partial}{\partial x_j} \frac{q_m(\iota - s, x)}{\rho_n(\iota - s, x)}]$$

$$= \int ds \int dx \, \kappa(s, x) \, [\, \frac{1}{\rho_n} \{ -\frac{\partial}{\partial s} q_m + \frac{1}{2} \nabla(\alpha \nabla q_m) - \nabla(\{\beta + \alpha \frac{\nabla^T \Phi_n}{\Phi_n}\} q_m) \}$$

$$- \frac{q_m}{\rho_n^2} \{ -\frac{\partial}{\partial s} \rho_n + \frac{1}{2} \nabla(\alpha \nabla \rho_n) - \nabla(\{\beta + \alpha \frac{\nabla^T \Phi_n}{\Phi_n}\} \rho_n) \}]$$

$$= \int ds \int dx \, q_m(s, x) \, \{ L(\frac{\kappa}{\rho_n}) + \frac{\nabla \Phi_n}{\Phi_n} \alpha \nabla^T \frac{\kappa}{\rho_n} \}(s, x)$$

$$- \int ds \int dx \, \rho_n(s, x) \, \{ L(\frac{\kappa q_m}{\rho_n^2}) + \frac{\nabla \Phi_n}{\Phi_n} \alpha \nabla^T \frac{\kappa q_m}{\rho_n^2} \}(s, x)$$

where $\kappa(s, x) = -\gamma(\iota - s, x)$. Since $q(v, z; ., .)$ and ρ are weak solutions of (2.79), the right-hand side of the last equality vanishes as m and n tend to infinity. The involved limit considerations are all managed similarly as in the proof of Lemma 2.2.2, hence they are omitted here. $\diamond$

Chapter 3

Integral and Diffusion Equations

Let us investigate the correspondence of weak solutions of parabolic differential equations with singular creation and killing and solutions of Feynman-Kac integral equations with locally integrable potential. The correspondence is stated in Section 3.4 where the solutions are treated as locally integrable functions, continuity is neither required nor established. In Section 3.2 we meet an integral equation with a solution which can be given in terms of the Feynman-Kac formula. Stochastic calculus leads in Section 3.3 to a 'killed' integral equation which provides a refined uniqueness condition of solutions.

3.1 Generators and transition densities

Classically, diffusion processes are accompanied by their semigroup and their generator. Under sufficiently regular circumstances, the transition density of the diffusion process is the kernel of the semigroup and the fundamental solution of the diffusion (Fokker-Planck) equation associated with the generator.

Our situation motivated from the theory of Schrödinger equations does typically not provide sufficiently regular circumstances. In fact, diffusion equations with in general singular potential c occur as deduced in Section 2.2. Their solutions were proved to exist in the sense of distributions induced by the local space-time Sobolev space $H_{loc}^{1,2}$. In Theorem 3.4.1 we will show an equivalence of diffusion equations $Lp + c\,p = 0$ with singular creation and killing c and so-called *Feynman-Kac integral equations* (3.6). As a consequence, we can concentrate our investigations on Feynman-Kac integral equations which turn out to be more successfully to treat in case of modest regularity circumstances. In particular, their uniqueness of solutions in case of $c = -L\Phi/\Phi$, Φ a Schrödinger multiplier in Definition 2.2.2 and L in (2.1), provides a crucial criterion for $P_{(s,x)}[N_s^b] = 1$ with $(s,x) \in D(\Phi)$. Following Corollary 2.3.2, this criterion causes diffusions with in general singular drift $\alpha \nabla \log \Phi$ to be conservative. In order to weaken the required integrability assumptions furthermore, we consider simultaneously the 'killed' integral equation (3.17) in Proposition 3.3.1.

The Feynman-Kac integral equation in (3.6) has a kernel $g^o(s,x;t,y)$ related to the diffusion process with generator $L + c\,I$, where the Borel-measurable function c may be extended real-valued. In order to ensure a sufficiently regular underlying diffusion process associated with L, we want to assume that the coefficients $\alpha(s,x) \in \mathbb{R}^{d^2}$ and $\beta(s,x) \in \mathbb{R}^d$ as well as their spatial derivatives $\partial_{x_i}\alpha(s,x)$, $\partial_{x_i}\partial_{x_j}\alpha(s,x)$ and $\partial_{x_i}\beta(s,x)$, $1 \leq i,j \leq d$, are bounded and jointly continuous in $(s,x) \in [a,b] \times \mathbb{R}^d$ as well as uniformly δ-Hölder continuous in $x \in \mathbb{R}^d$ for a $\delta \in (0,1]$. In addition, the diffusion matrix α is required to be uniformly Hölder continuous in $s \in [a,b]$ and to satisfy the uniform ellipticity condition

$$\sum_{i,j=1}^{d} \alpha_{ij}(t,x)\, y_i\, y_j \geq \kappa \sum_{i=1}^{d} y_i^2, \quad \forall y \in \mathbb{R}^d$$

for a positive constant κ. As a consequence, theorems on parabolic differential equations due to S. Itô (1957) and Il'in, Kalashnikov and Oleinik (Dynkin (1965) vol. II appendix § 6) provide a *fundamental solution* $g(s,x;t,y)$ of (3.1) on $[a,b] \times \mathbb{R}^d$, i.e., $g(s,x;t,y)$ satisfies

$$L\,p = 0 \tag{3.1}$$

as a function of (s,x) and the adjoint equation

$$L^*\,p = 0 \tag{3.2}$$

as a function of (t,y), respectively, where L is given in (2.1). Following Dynkin (1965) vol. I chap. V §6, $g(s,x;t,y)$, $a \leq s \leq t \leq b$, $x,y \in \mathbb{R}^d$, can be considered as the transition probability density of a time-inhomogeneous diffusion process with space-time generator L. Due to Stroock-Varadhan (1979)'s theorem 5.2.2, there exists a non-negative definite symmetric square root σ of α. Kunita (1990)'s corollary 4.2.7 yields a Brownian flow $\xi_{s,t}$ required in Section 4.3 whose 1-point motion $\xi_{s,t}(x)$, $s \le t \le b$, $x \in \mathbb{R}^d$ fixed, has the space-time generator L. $\xi_{s,t}$ is described in Definition 2.3.1 by a family of measures $P_{(s,x)}$ on a space of continuous paths and it possesses the transition probability density $g(s,x;t,y)$ where the identification follows from Stroock-Varadhan (1979)'s uniqueness theorem 3.2.6. Their § 5.2 implies that the coefficients σ and β satisfy the assumptions $(\mathcal{A})$ in Section 4.3 with the Hölder exponent $\delta \in (0, 1]$ provided above. Hence $\xi_{s,t}$ defined in (4.10) is a forward $C^{1,\varepsilon}$-Brownian flow of local C^1-diffeomorphisms for $0 < \varepsilon < \delta$, which will be essential in Chapter 4.

3.2 Feynman-Kac integral equations

Let c be an extended real-valued Borel-measurable function on $[a, b] \times \mathbb{R}^d$. We set

$$D = \{(s, x) \in [a, b] \times \mathbb{R}^d : |c(s, x)| < \infty\} \qquad (3.3)$$

and assume that $D \cap (a, b) \times \mathbb{R}^d$ is an open subset of $\mathbb{R}^{d+1}$. Let us denote the first exit time of $((t, X_t), P_{(s,x)})$ in Definition 2.3.1 from D by

$$\tau_s = \inf\{u \in [s, b) : (u, X_u) \notin D\}. \qquad (3.4)$$

It is our intention to characterize weak solutions p of $Lp + cp = 0$ on D, i.e., locally integrable functions p which satisfy

$$\int_D ds\, dx\, \{L^*(s, x)\gamma(s, x) + c(s, x)\, \gamma(s, x)\}\, p(s, x) = 0$$

for all $\gamma \in C^\infty_{comp}(D)$.

Let us define

$$g^o(s, x; t, y) = g(s, x; t, y) - P_{(s,x)}[g(\tau_s, X_{\tau_s}; t, y); t > \tau_s] \qquad (3.5)$$

for $a \le s < t \le b$ and $x, y \in \mathbb{R}^d$, where $g(s, x; t, y) = 0$ if $(s, x) \notin D$. What we consider in (3.5) is the transition density of traveling from (s, x) to (t, y) without leaving D. It satisfies

$$P_{(s,x)}[f(t, X_t); t < \tau_s] = \int g^o(s, x; t, y)\, dy\, f(t, y)$$

for bounded and measurable functions f on $[a, b] \times I\!\!R^d$ with supp$f \subset D$. In fact, the left-hand side equals

$$P_{(s,x)}[f(t, X_t)] - P_{(s,x)}[f(t, X_t); t > \tau_s] - P_{(s,x)}[f(t, X_t); t = \tau_s]$$

where the second term becomes

$$P_{(s,x)}[P_{(\tau_s, X_{\tau_s})}[f(t, X_t); t > \tau_s]] = \int P_{(s,x)}[g(\tau_s, X_{\tau_s}; t, y); t > \tau_s]\, f(t, y)\, dy$$

by means of the Markov property and the third term vanishes because of supp $f \subset D$.

Lemma 3.2.1 *The transition density $g^o(s, x; t, y)$ defined in (3.5) satisfies the diffusion equation (3.2) as a function of (t, y).*

Proof. Let $g(t, y)$ denote $g(\tau_s, X_{\tau_s}; t, y)$. Then

$$\frac{1}{h}\{P_{(s,x)}[g(t + h, y); t + h > \tau_s] - P_{(s,x)}[g(t, y); t > \tau_s]\}$$

$$= \frac{1}{h} P_{(s,x)}[g(t + h, y); t + h > \tau_s \geq t] + \frac{1}{h} P_{(s,x)}[g(t + h, y) - g(t, y); t > \tau_s]$$

where the first member of the right-hand side vanishes as $h \searrow 0$. In fact, the continuity of paths yields

$$\frac{1}{h} P_{(\tau_s, X_{\tau_s})}[X_{t+h} \in U] \to 0$$

on $\{t + h > \tau_s \geq t\}$, $P_{(s,x)}$-a.s., where U is a neighborhood of y such that $[t, t + h] \times U \subset D$. The second member of the right-hand side converges to

$$P_{(s,x)}[\frac{\partial}{\partial t} g(t, y); t > \tau_s]$$

as $h \searrow 0$. Hence, both members of the right-hand side of (3.5) satisfy (3.2) as functions of (t, y). $\diamond$

We are going to investigate the so-called Feynman-Kac integral equation (3.6) under mild assumptions on the function c which will be satisfied by the weak reference potential in (2.22). Let us consider

$$\begin{aligned} p(s, x) &= \int g^o(s, x; t, y)\, dy\, f(t, y) \\ &+ \int_s^t dv \int g^o(s, x; v, y)\, dy\, c(v, y)\, p(v, y) \end{aligned} \tag{3.6}$$

for $(s, x) \in D$, $a \leq s < t$, $t \in (a, b]$ fixed, under the integrability conditions

$$\int g^\circ(s, x; v, y) \, dy \mid p(v, y) \mid < \infty \tag{3.7}$$

for $(s, x) \in D$, $a \leq s < v \leq t$, and

$$\int_s^t dv \int g^\circ(s, x; v, y) \, dy \mid c(v, y) \, p(v, y) \mid < \infty \tag{3.8}$$

for $(s, x) \in D$, $a \leq s < t$. We notice that any solution p of (3.6) satisfies

$$p(s, x) \;=\; \int g^\circ(s, x; u, y) \, dy \, p(u, y) \tag{3.9}$$
$$+ \; \int_s^u dv \int g^\circ(s, x; v, y) \, dy \, c(v, y) \, p(v, y)$$

for $(s, x) \in D$, $a \leq s < u \leq t$. In fact,

$$\int g^\circ(s, x; u, y) \, dy \, p(u, y)$$
$$= \int g^\circ(s, x; u, y) \, dy \int g^\circ(u, y; t, z) \, dz \, f(t, z)$$
$$+ \int g^\circ(s, x; u, y) \, dy \int_u^t dv \int g^\circ(u, y; v, z) \, dz \, c(v, z) \, p(v, z)$$
$$= \int g^\circ(s, x; t, z) \, dz \, f(t, z) + \int_u^t dv \int g^\circ(s, x; v, z) \, dz \, c(v, z) \, p(v, z)$$

which is the first member of the right-hand side of (3.9). Hence its left-hand side is computed as

$$\int g^\circ(s, x; u, y) \, dy \, p(u, y) + \int_s^u dv \int g^\circ(s, x; v, y) \, dy \, c(v, y) \, p(v, y)$$
$$= \int g^\circ(s, x; t, z) \, dz \, f(t, z) + \int_u^t dv \int g^\circ(s, x; v, z) \, dz \, c(v, z) \, p(v, z)$$
$$+ \int_s^u dv \int g^\circ(s, x; v, y) \, dy \, c(v, y) \, p(v, y) = p(s, x)$$

by means of (3.6). As a consequence, condition (3.7) may be replaced by

$$\int g^\circ(s, x; t, y) \, dy \, f(t, y) < \infty \tag{3.10}$$

for $(s, x) \in D$, $a \leq s < t$, in case of non-negative solutions p of (3.6) which satisfy (3.8).

Proposition 3.2.1 (Nagasawa (1989)'s theorem 3.2) *Let the Feynman-Kac integral equation (3.6) possess a potential c which is continuous on D.*

(i) Let $f(t, .)$ be a non-negative function with (3.10) such that

$$\tilde{p}(s, x; t, f) = P_{(s,x)}[f(t, X_t) \exp\{\int_s^t c(v, X_v)\, dv\}; t < \tau_s] \tag{3.11}$$

satisfies (3.8). Then, $\tilde{p}(s, x; t, f)$ is a solution of (3.6).

(ii) Let $f(t, .)$ with

$$\int g^o(s, x; t, y)\, dy \mid f(t, y) \mid < \infty \tag{3.12}$$

for $(s, x) \in D$, $a \le s < t$, such that $\tilde{p}(s, x; t, \mid f \mid)$ satisfies (3.8). Then, $\tilde{p}(s, x; t, f)$ is a solution of (3.6).

Proof. (i) We have already remarked that (3.10) is enough for a non-negative solution p of (3.6) to satisfy (3.7). Under (3.8), the second member of the right-hand side of (3.6) is well-defined and with (3.11) in place of p, the Markov property and Fubini's theorem turn it into

$$\int_s^t du\, P_{(s,x)}[c(u, X_u)$$

$$\cdot P_{(u,X_u)}[f(t, X_t) \exp\{\int_u^t c(v, X_v)\, dv\}; t < \tau_u]; u < \tau_s]$$

$$= \int_s^t du\, P_{(s,x)}[c(u, X_u) \exp\{\int_u^t c(v, X_v)\, dv\}\, f(t, X_t); t < \tau_s]$$

$$= P_{(s,x)}[\int_s^t du\, c(u, X_u) \sum_{n=1}^{\infty} \frac{1}{(n-1)!} (\int_u^t c(v, X_v)\, dv)^{n-1}\, f(t, X_t); t < \tau_s]$$

$$= P_{(s,x)}[\sum_{i=1}^{\infty} \frac{1}{n!} (\int_s^t c(v, X_v)\, dv)^n\, f(t, X_t); t < \tau_s].$$

Hence, completed by its first member, the right-hand side of (3.6) equals

$$P_{(s,x)}[\sum_{i=0}^{\infty} \frac{1}{n!} (\int_s^t c(v, X_v)\, dv)^n\, f(t, X_t); t < \tau_s]$$

which is just $\tilde{p}(s, x; t, f)$, i.e., the claimed left-hand side of (3.6).

(*ii*) Under the stated conditions, $\tilde{p}(s, x; t, |\ f\ |)$ is a solution of (3.6) with $|\ f\ |$ in place of f by part (*i*). We notice that (3.7) is

$$\int g^o(s, x; u, y)\, dy\, \tilde{p}(u, y; t, |\ f\ |)\ < \infty$$

for $(s, x) \in D$, $s < u \le t$, and that

$$|\ \tilde{p}(s, x; t, f)\ | \le\ \tilde{p}(s, x; t, |\ f\ |).$$

Hence $\tilde{p}(s, x; t, f)$ is well-defined and satisfies (3.6), (3.7) and (3.8). $\Diamond$

Proposition 3.2.2 (Nagasawa (1989)'s theorem 3.1) *Under the integrability conditions (3.7) and*

$$\int_s^t du\, P_{(s,x)}[\exp\{\int_s^u |\ c(v, X_v)\ |\ dv\}\ |\ c\, p(u, X_u)\ |; u < \tau_s] < \infty \qquad (3.13)$$

the integral equation (3.6) has a unique solution provided the potential function c is continuous on D.

Proof. We first notice that (3.13) implies (3.8). Let $p_1(s, x)$ and $p_2(s, x)$ be two solutions of (3.6) with (3.7) and (3.13). Then $d(s, x) = |\ p_1(s, x) - p_2(s, x)\ |$ can be estimated by

$$d(s, x) \le \int_s^t dv\, P_{(s,x)}[|\ c(v, X_v)\ |\ d(v, X_v); v < \tau_s]$$

$$\le \int_s^t dv\, P_{(s,x)}[|\ c(v, X_v)\ |\ \int_v^t dv_1\, P_{(v, X_v)}[|\ c\ |\ d(v_1, X_{v_1}); v_1 < \tau_v]; v < \tau_s]$$

$$\le \int_s^t dv_1 \int_s^{v_1} dv\, P_{(s,x)}[|\ c(v, X_v)\, c(v_1, X_{v_1})\ |\ d(v_1, X_{v_1}); v_1 < \tau_s]$$

where the Markov property and Fubini's theorem have been employed. Inductive repetition yields

$$d(s, x) \le \int_s^t dv_n \int_s^{v_n} dv_{n-1} \cdots \int_s^{v_1} dv$$
$$\cdot P_{(s,x)}[|\ c(v, X_v)\, c(v_1, X_{v_1}) \cdot \ldots \cdot c(v_n, X_{v_n})\ |\ d(v_n, X_{v_n}); v_n < \tau_s]$$

$$\le \int_s^t dv\, P_{(s,x)}[\frac{1}{(n-1)!}\, (\int_s^v |\ c(u, X_u)\ |\ du)^{n-1}\ |\ c\, d(v, X_v)\ |; v < \tau_s]$$

which vanishes as n tends to infinity because of (3.13). $\Diamond$

Lemma 3.2.2 *Let $c \in L^1_{loc}(D)$. For $(s,x) \in D$ and $s < t \le b$*

$$\frac{1}{n} \left(\int_s^t c(u, X_u)\, du \right)^n = \int_s^t dv \left(\int_s^v c(u, X_u)\, du \right)^{n-1} c(v, X_v) \quad (3.14)$$

$$= \int_s^t dv \left(\int_v^t c(u, X_u)\, du \right)^{n-1} c(v, X_v)$$

on $\{t < \tau_s\}$, $P_{(s,x)}$-a.s.

Proof. Let us refer to Definition 2.3.1 and Lemma 4.3.1. For P-almost all $\omega \in \{t < \tau_{(s,x)}\}$, the mapping

$$[s, t] \to D, \ v \mapsto (v, \xi^x_{s,v}(\omega))$$

is continuous and hence its range is compact. Consequently, there exists a sequence $(c_{\omega,m})_{m \in \mathbb{N}}$ of continuous functions such that

$$L^1([s,t]) - \lim_{m \to \infty} c_{\omega,m} = c(\,.\,, \xi^x_{s,\,.}(\omega)).$$

In case of continuous integrands like $c_{\omega,m}$, $m \in \mathbb{N}$, (3.14) is verified by taking derivatives with respect to the upper and lower bound, respectively. Thus, the case $c \in L^1_{loc}(D)$ follows from

$$\left| \int_s^t dv \left(\int_s^v c_{\omega,m}(u)\, du \right)^{n-1} c_{\omega,m}(v) \right.$$

$$\left. - \int_s^t dv \left(\int_s^v c(u, \xi^x_{s,u}(\omega))\, du \right)^{n-1} c(v, \xi^x_{s,v}(\omega)) \right|$$

$$\le \int_s^t dv \, \left| c_{\omega,m}(v) - c(v, \xi^x_{s,v}(\omega)) \right| \, \left| \int_s^v c_{\omega,m}(u)\, du \right|^{n-1}$$

$$+ \left| \int_s^t dv \, c(v, \xi^x_{s,v}(\omega)) \right|$$

$$\cdot \left| \left(\int_s^v c_{\omega,m}(u)\, du \right)^{n-1} - \left(\int_s^v c(u, \xi^x_{s,u}(\omega))\, du \right)^{n-1} \right| \to 0$$

as $m \nearrow \infty$. $\Diamond$

The above results immediately yield

Corollary 3.2.1 *Proposition 3.2.1 and Proposition 3.2.2 hold analogously for potential functions $c \in L^1_{loc}(D)$.*

Lemma 3.2.3 *Let h_1 and h_2 be functions defined on D which satisfy (3.7) and (3.8) with h_2 in place of $c\,p$, respectively. Then*

$$h_1,\ h_2 \in L^1_{loc}(\{(v,y) \in D,\ v \in (a,t]\}).$$

Proof. Let V be any compact set contained in $\{(v,y) \in D : v \in (a,t]\}$. There exists $(s,x) \in D\backslash V$ in the same connected component of D as V such that

$$
\begin{aligned}
s \quad &< \quad v_1 = \min\{v \in (a,b) : \exists\, y \in I\!\!R^d \text{ with } (v,y) \in V\} \\
&\leq \quad v_2 = \max\{v \in (a,b) : \exists\, y \in I\!\!R^d \text{ with } (v,y) \in V\} \ \leq t.
\end{aligned}
$$

Lemma 3.2.1 claims that $g^o(s,x;v,y)$ is jointly continuous in (v,y). Hence there exists $\eta > 0$ such that $\eta \leq g^o(s,x;v,y)$ for all $(v,y) \in V$ and

$$\int_V\!\!\int |\,h_2(v,y)\,|\ dv\,dy \leq \frac{1}{\eta} \int_{v_1}^{v_2} dv \int dy\, g^o(s,x;v,y)\ |\,h_2(v,y)\,|\ < \infty$$

as well as

$$\int_V\!\!\int |\,h_1(v,y)\,|\ dv\,dy \leq \frac{v_2 - v_1}{\eta} \sup_{v\in[v_1,v_2]} \int g^o(s,x;v,y)\,dy\ |\,h_1(v,y)\,|\ < \infty$$

follow. $\Diamond$

Lemma 3.2.4 (Nagasawa (1989)'s lemma 3.1) *Let p be a function on D which satisfies (3.8). Then, for any $\varepsilon > 0$, there exist $\delta_0 > 0$ and an open subset U of D with compact closure $\overline{U} \subset D$ such that $(v,x) \in U$ for $v \in (s - \delta_0, s]$ and*

$$\frac{1}{\delta} \int_{s-\delta}^{s} dv \int g^o(s-\delta,x;v,y)\,dy\ |\,c\,p(v,y)\,|\ 1_{U^c}(v,y) < \varepsilon \tag{3.15}$$

for $0 < \delta < \delta_0$.

Proof. Let $(s,x) \in D$ and let $\delta_0 > 0$ so small that $(v,x) \in D$ for all $v \in [s-\delta_0, s]$. We define a measure μ_δ for $0 < \delta < \delta_0$ by

$$\mu_\delta(dv\,dy) = \frac{1}{\delta}\, g^o(s-\delta,x;v,y)\, 1_{(s-\delta,s]}(v)\, dv\,dy$$

and notice that

$$\mu(D) = \int_a^b dv\, \frac{1}{\delta}\, 1_{(s-\delta,s]}(v) \int_{I\!\!R^d} g^o(s-\delta,x;v,y)\,dy \leq 1.$$

Since $cp \in L^1(\mu_{\delta_0})$ because of (3.8), there exists an open subset U of D with the compact closure $\overline{U} \subset D$ such that $(v, x) \in U$ for $v \in (s - \delta_0, s]$ and

$$\int \int \mu_{\delta_0}(dv\, dy) \mid c(v, y)\, p(v, y) \mid 1_{U^c}(v, y) < \varepsilon.$$

Hence, by the continuity of paths of the process $(X_t, P_{(s,x)})$,

$$\mu_\delta(U^c) \quad - \quad \int_a^b dv \int_{\mathbb{R}^d} dy\, \frac{1}{\delta} g^\upsilon(s - \delta, x; v, y)\, 1_{(s-\delta, s]}(v)\, 1_{U^c}(v, y)$$

$$= \quad P_{(s-\delta, x)}[\int_a^b dv\, \frac{1}{\delta} 1_{(s-\delta, s]}(v)\, 1_{U^c}(v, X_v); v < \tau_{s-\delta}] \searrow 0$$

as $\delta \searrow 0$ which implies (3.15). $\diamond$

Lemma 3.2.5 *Let K be a compact subset of $\mathbb{R}^d$ such that there exist $s_1 \in (a, b)$ and $\delta_1 > 0$ with*

$$\{(s, x) : s \in [s_1 - 2\delta_1, s_1], \ x \in K\} \subset D.$$

Let p be a function on D which satisfies (3.8). Then, for any $\varepsilon > 0$, there exist $\delta_0 > 0$ and a jointly continuous function j_ε with compact support contained in D such that

$$\frac{1}{\delta} \int_{s-\delta}^s dv \int g^o(s - \delta, x; v, y)\, dy\, c(v, y)\, p(v, y)$$

$$= \frac{1}{\delta} \int_{s-\delta}^s dv \int g^o(s - \delta, x; v, y)\, dy\, c(v, y)\, p(v, y)\, j_\varepsilon(v, y) + O(\varepsilon, \delta_0)$$

for $(s, x) \in [s_1 - \delta_1, s_1] \times K$ where $O(\varepsilon, \delta_0) \to 0$ as $\varepsilon \searrow 0$ uniformly on the set $[s_1 - \delta_1, s_1] \times K$ for $0 < \delta < \frac{1}{2}\delta_0$.

Proof. Lemma 3.2.4 provides for any $\varepsilon > 0$ and for any $(s_0, x_0) \in D$ a small $\delta_0(s_0, x_0) > 0$ and two open neighborhoods $U(s_0, x_0)$, $V(s_0, x_0)$ with compact closures contained in D such that

$$\frac{1}{\delta} \int_{s-\delta}^s dv \int g^o(s - \delta, x; v, y)\, dy \mid c(v, y)\, p(v, y) \mid 1_{U^c(s_0, x_0)}(v, y) < 2\varepsilon \quad (3.16)$$

for $(s, x) \in V(s_0, x_0)$ where $0 < \delta < \frac{1}{2}\delta_0(s_0, x_0)$. Applied for fixed $\varepsilon > 0$ along a dense sequence $((s_n, x_n))_{n \geq 2}$ in the compact set $[s_1 - \delta_1, s_1] \times K$, we obtain $(\delta_0(s_n, x_n))_{n \geq 2}$ and two sequences of neighborhoods $(V(s_n, x_n))_{n \geq 2}$

and $(U(s_n, x_n))_{n \geq 2}$. As a consequence, there exists a finite set $J \subset I\!N$ such that

$$[s_1 - \delta_1, s_1] \times K \subset \bigcup_{j \in J} V(s_j, x_j).$$

In case of

$$\delta_0 = \min_{j \in J} \delta_0(s_j, x_j) \wedge \delta_1 > 0$$

and

$$U = \bigcup_{j \in J} U(s_j, x_j) \cup V(s_j, x_j)$$

relation (3.16) holds for all $(s, x) \in [s_1 - \delta_1, s_1] \times K \subset U \subset \overline{U} \subset D$ where $\overline{U}$ is compact. Moreover, there exists an open subset U_1 with compact closure such that $\overline{U} \subset U_1 \subset \overline{U_1} \subset D$. Hence j_ε can be any jointly continuous function on D which satisfies

$$
\begin{aligned}
j_\varepsilon(v, y) &\in [0, 1] & \text{for } (v, y) \in D \\
j_\varepsilon(v, y) &= 1 & \text{for } (v, y) \in U \\
j_\varepsilon(v, y) &= 0 & \text{for } (v, y) \in U_1^c.
\end{aligned}
$$

$\Diamond$

3.3 'Killed' integral equations

The uniqueness in case of Feynman-Kac integral equations (3.6) can be considered separately for non-negative solutions which additionally satisfy weakly $Lp + cp = 0$. T. Sturm suggested to involve a killing c^- which yields the uniqueness condition (3.19) appearing in Nagasawa (1990)'s theorem 1. We want to extend (3.19) to non-smooth solutions p including the situation of Schrödinger multipliers in Definition 2.2.2 which was treated in Corollary 2.3.2.

Proposition 3.3.1 will be proved essentially by Itô's formula for non-smooth functions in Section 4.3. Hence let p be a non-negative, extended real-valued, measurable function on $[a, b] \times I\!R^d$ with derivatives in the sense of distributions which satisfy

$$D_s p \in L^1_{loc}(\{p > 0\}), \quad D_i p \in L^2_{loc}(\{p > 0\}) \quad \text{and} \quad D_i D_j p \in L^1_{loc}(\{p > 0\})$$

for $1 \leq i, j \leq d$. We assume that the subset

$$
\begin{aligned}
D(p) \quad = \quad & \{(s, x) \in \{p > 0\} : p \text{ is continuous at } (s, x) \text{ and} \\
& \mid D_s p(s, x) \mid < \infty, \mid D_i p(s, x) \mid < \infty \text{ as well as} \\
& \mid D_i D_j p(s, x) \mid < \infty \text{ for } 1 \leq i, j \leq d \}
\end{aligned}
$$

is open and define the reference potential c associated with L in (2.1) and p by

$$
c(s, x) = -\frac{Lp(s, x)}{p(s, x)} \quad \text{for } (s, x) \in D(p).
$$

Hence $c \in L^1_{loc}(D(p))$ where the positive and negative part of c will be denoted by $c^+ = c \vee 0$ and $c^- = -c \vee 0$, respectively.

Let $((t, X_t), \, P_{(s,x)}, \, (s, x) \in [a, b] \times I\!\!R^d, \, s \leq t \leq b)$ be the space-time diffusion process in Definition 2.3.1 with generator L in (2.1) employed in (3.1). Following Lemma 4.3.1, path integrals of c along $((t, X_t), \, P_{(s,x)})$ are well-defined on $D(p)$ for $a \leq s \leq t$ and λ^d-a.a. $x \in I\!\!R^d$. Hence

$$
\tau_s = \inf\{u \in [s, b) : (u, X_u) \notin D(p)\}
$$

allows us to consider for $(s, x) \in D(p)$ the killed process

$$
P^-_{(s,x)}[F; t < \tau_s] = P_{(s,x)}[\exp\{-\int_s^t c^-(v, X_v) \, dv\} F; t < \tau_s]
$$

for bounded, $\sigma((v, X_v), \, s \leq v \leq t)$-measurable functions F on $D(p)$.

Proposition 3.3.1 *Let*

$$
p(s, x) = P^-_{(s,x)}[p(t, X_t); t < \tau_s] + \int_s^t du \, P^-_{(s,x)}[c^+ p(u, X_u); u < \tau_s] \quad (3.17)
$$

for $(s, x) \in D(p)$ with $a \leq s \leq t$ be called 'killed' integral equation. Let

$$
P_{(s,x)}[\exp\{\int_s^t c(v, X_v) \, dv\} p(t, X_t); t < \tau_s] < \infty \quad (3.18)
$$

and

$$
\int_s^t du \, P_{(s,x)}[\exp\{\int_s^u c(v, X_v) \, dv\} c^+ p(u, X_u); u < \tau_s] < \infty \quad (3.19)
$$

be two integrability conditions, where $t \in (a, b]$ is fixed. If any non-negative function p satisfies the assumptions of Itô's formula in Theorem 4.3.1, the *integrability conditions (3.18) and (3.19) as well as $Lp + cp = 0$ weakly and if $p(v, \, . \,)$ is continuous at time $t \in (a, b]$, then p solves the 'killed' integral equation (3.17) for $a \leq s \leq t$ and λ^d-a.a. $x \in \mathbb{R}^d$ with $(s, x) \in D(p)$. Since the 'killed' integral equation (3.17) has a unique solution among such functions p, they all coincide for $a \leq s \leq t$, λ^d-a.e. on $\{x \in \mathbb{R}^d : (s, x) \in D(p)\}$.*

Proof. In the first step we show that these functions p are solutions of (3.17). The verification essentially consists of an application of Itô's formula in Section 4.3 to the product of non-smooth functions

$$p(t, X_t) \exp\{- \int_s^t c^-(u, X_u) \, du\} \quad \text{on} \quad \{t < \tau_s\}$$

generated along $(\,(t, X_t), \, P_{(s,x)}^-\,)$. Thus, taking the expectation w.r.t. $P_{(s,x)}^-$ will yield (3.17). In order to employ Theorem 4.3.1, let $(D^{(n)})_{n \in \mathbb{N}}$ be an increasing sequence of open sets with compact closures $\overline{D^{(n)}}$ contained in $D(p)$ such that

$$D(p) = \bigcup_{n \in \mathbb{N}} D^{(n)}.$$

The associated first exit times are given by

$$\tau_s^{(n)} = \inf\{u \in [s, b) : (u, X_u) \notin D^{(n)}\}$$

where $\tau_s^{(n)} \leq \tau_s$, $P_{(s,x)}$-a.s. for $n \in \mathbb{N}$, and $\lim_{n \nearrow \infty} \tau_s^{(n)} = \tau_s$ on $\{\tau_s < \infty\}$, $P_{(s,x)}$-a.s. For

$$G_t = p(t, X_t) \quad \text{and} \quad H_t = \exp\{h_t\}$$

with

$$h_t = - \int_s^t c^-(v, X_v) \, dv$$

on $\{t < \tau_s^{(n)}\}$ we receive

$$dG_t = D_s p \, dt + \nabla p \, dX_t + \frac{1}{2} \triangle_\alpha p \, dt$$

and

$$dH_t = (e^h)' \, dh_t + \frac{1}{2}(e^h)'' \, d < h >_t = -c^-(t, X_t) \, H_t \, dt$$

on $\{t < \tau_s^{(n)}\}$, $P_{(s,x)}$-a.s. for λ^d-a.e. on $\{x \in \mathbb{R}^d : (s, x) \in D(p)\}$ and $n \in \mathbb{N}$. Let us recall that $(dB_t)^2 = dt$, $dB_t \, dt = 0$ and $(dt)^2 = 0$ where $(B_t)_{s \leq t \leq b}$ is the

Brownian motion w.r.t. $P_{(s,x)}$. Applied along $dX_t = \sigma(t, X_t)\, dB_t + \beta(t, X_t)\, dt$ which was treated in Section 3.1, the stochastic calculus yields

$$
\begin{aligned}
& d(p(t, X_t)\, \exp\{-\int_s^t c^-(v, X_v)\, dv\})\, 1_{\{t < \tau_s^{(n)}\}} \\
= \; & d(G_t\, H_t)\, 1_{\{t < \tau_s^{(n)}\}} = (dG_t\, H_t + G_t\, dH_t + dG_t\, dH_t)\, 1_{\{t < \tau_s^{(n)}\}} \\
= \; & -c^-(t, X_t)\, p(t, X_t)\, \exp\{-\int_s^t c^-(v, X_v)\, dv\}\, 1_{\{t < \tau_s^{(n)}\}}\, dt \\
+ \; & [D_s p(t, X_t)\, dt + \nabla p(t, X_t)\, dX_t + \frac{1}{2}\, \triangle_\alpha\, p(t, X_t)\, dt] \\
& \cdot \exp\{-\int_s^t c^-(v, X_v)\, dv\}\, 1_{\{t < \tau_s^{(n)}\}} \\
- \; & c^-(t, X_t)\, \exp\{-\int_s^t c^-(v, X_v)\, dv\}\, 1_{\{t < \tau_s^{(n)}\}}\, dt \\
& \cdot [D_s p(t, X_t)\, dt + \nabla p(t, X_t)\, (\sigma(t, X_t)\, dB_t + \beta(t, X_t)\, dt) \\
& + \frac{1}{2}\, \triangle_\alpha\, p(t, X_t)\, dt] \\
= \; & -c^-(t, X_t)\, p(t, X_t)\, \exp\{-\int_s^t c^-(v, X_v)\, dv\}\, 1_{\{t < \tau_s^{(n)}\}}\, dt \\
+ \; & Lp(t, X_t)\, \exp\{-\int_s^t c^-(v, X_v)\, dv\}\, 1_{\{t < \tau_s^{(n)}\}}\, dt \\
+ \; & \nabla p(t, X_t)\, \sigma(t, X_t)\, \exp\{-\int_s^t c^-(v, X_v)\, dv\}\, 1_{\{t < \tau_s^{(n)}\}}\, dB_t \\
= \; & -c^+(t, X_t)\, p(t, X_t)\, \exp\{-\int_s^t c^-(v, X_v)\, dv\}\, 1_{\{t < \tau_s^{(n)}\}}\, dt \\
+ \; & \nabla p(t, X_t)\, \sigma(t, X_t)\, \exp\{-\int_s^t c^-(v, X_v)\, dv\}\, 1_{\{t < \tau_s^{(n)}\}}\, dB_t
\end{aligned}
$$

$P_{(s,x)}$-a.s. for λ^d-a.a. $x \in \mathbb{R}^d$ with $(s, x) \in D^{(n)}$ and $n \in \mathbb{N}$, where $Lp = -cp$ and $cp = c^+ p - c^- p$ have been employed. By pathwise integration we receive

$$
\begin{aligned}
p(s, x) \; = \; & p(t, X_t)\, \exp\{-\int_s^t c^-(v, X_v)\, dv\}\, 1_{\{t < \tau_s^{(n)}\}} \\
+ \; & \int_s^t \exp\{-\int_s^v c^-(u, X_u)\, du\}\, c^+(v, X_v)\, p(v, X_v)\, 1_{\{v < \tau_s^{(n)}\}}\, dv \\
+ \; & \int_s^t \exp\{-\int_s^v c^-(u, X_u)\, du\}\, \nabla p(v, X_v)\, \sigma(v, X_v)\, 1_{\{v < \tau_s^{(n)}\}}\, dB_v
\end{aligned}
$$

$P_{(s,x)}$-a.s. for λ^d-a.a. $x \in \mathbb{R}^d$ with $(s, x) \in D^{(n)}$ and $n \in \mathbb{N}$. Taking the expectation with respect to $P_{(s,x)}$ yields

$$p(s,x) = P_{(s,x)}[p(t,X_t)\exp\{-\int_s^t c^-(v,X_v)\,dv\}; t < \tau_s^{(n)}]$$

$$+ \quad P_{(s,x)}[\int_s^t dv\, c^+(v,X_v)\, p(v,X_v)\exp\{-\int_s^v c^-(u,X_u)\,du\}; t < \tau_s^{(n)}]$$

$$+ \quad P_{(s,x)}[\int_s^t \nabla p(v,X_v)\,\sigma(v,X_v)\exp\{-\int_s^v c^-(u,X_u)\,du\}\,1_{\{t<\tau_s^{(n)}\}}\,dB_v]$$

for λ^d-a.a. $x \in \mathbb{R}^d$ with $(s,x) \in D^{(n)}$ and $n \in \mathbb{N}$, where the third member of the right-hand side vanishes. In fact, the stochastic integral is a martingale since its integrand is bounded on $D^{(n)}$ for $n \in \mathbb{N}$. In order to finally arrive at (3.17), we have to take the limit as n tends to infinity. In case of the first member of the right-hand side, the limit is a consequence of Beppo-Levi's theorem. In case of the second member of the right-hand side, the integrability condition (3.19) implies dominated convergence, hence the limit is provided by Lebesgue's theorem.

The second step consists of a procedure analogous to the proof of Proposition 3.2.1. Because of Lemma 3.2.2 we find under (3.18) that the function

$$p^o(s,x) = P_{(s,x)}[\exp\{\int_s^t c(v,X_v)\,dv\}\, p(t,X_t); t < \tau_s] \qquad (3.20)$$

is a solution of (3.17), $s < t$, with $p^o(t,x) = p(t,x)$. Hence $p^o(s,x)$ coincides with $\tilde{p}(s,x;t,f)$ in (3.11) in case of $f(t,\,.\,) = p(t,\,.\,)$.

In the third step we take two solutions $p_1(s,x)$ and $p_2(s,x)$ of the 'killed' integral equation (3.17) which satisfy the conditions (3.18) and (3.19). Their difference $d(s,x) =| p_1(s,x) - p_2(s,x) |$ satisfies analogously to the proof of Proposition 3.2.2

$$d(s,x) \le \int_s^t dv\, P_{(s,x)}^-[c^+\, d(v,X_v); v < \tau_s]$$

$$\le \int_s^t dv\, P_{(s,x)}^-[\frac{1}{(n-1)!}\,(\int_s^v c^+(u,X_u)\,du)^{n-1}\, c^+\, d(v,X_v); v < \tau_s]$$

for $a \le s \le t$ and λ^d-a.a. $x \in \mathbb{R}^d$ with $(s,x) \in D(p)$. Because of (3.19) the right-hand side vanishes as n tends to infinity. $\diamondsuit$

Remark 3.3.1 The integrability condition (3.18) implies

$$P_{(s,x)}^-[p(t,X_t); t < \tau_s] < \infty$$

and the integrability condition (3.19) implies

$$\int_s^t du\, P^-_{(s,x)}[c^+ p(u, X_u); u < \tau_s] < \infty.$$

We realize that the two implications are the integrability conditions for the 'killed' integral equation (3.17) corresponding to (3.7) and (3.8). Discussing the integrability assumptions in Proposition 3.3.1, we notice first that (3.18) is sufficient for the first member of the right-hand side of (3.17) to be well-defined and that (3.19) is sufficient for the second member of the right-hand side of (3.17) to be well-defined. More precisely, (3.18) and (3.19) with $-c^-$ in place of c would be sufficient for the 'killed' integral equation (3.17) to be well-defined. But (3.13) is needed for the particular solution $p^o(s, x)$ of (3.17) to be well-defined and (3.19) implies the uniqueness of solutions of (3.17) in the considered class of functions.

3.4   Equivalence of solutions

Theorem 3.4.1 *Let $p(s, x; t, f)$ be a solution of the Feynman-Kac integral equation (3.6) with (3.7) and (3.8) where $f(t, \,.\,)$ is bounded and continuous on $\{y : (t, y) \in D\}$. Then, $p(s, x; t, f)$ is differentiable in s, λ^{d+1}-almost everywhere on D, and a weak solution of the parabolic differential equation*

$$\frac{\partial}{\partial s} p + \frac{1}{2}\triangle_\alpha p + \beta \cdot \nabla p + cp \;=\; 0 \quad on\; D \tag{3.21}$$

satisfying the terminal condition

$$\lim_{s \nearrow t} p(s, x) \;=\; f(t, x) \quad for\; (t, x) \in D. \tag{3.22}$$

    Conversely, let p be a weak solution of (3.21) with (3.22) where $p(v, y)$ is left-continuous in v at final time t, λ^d-a.e. on $\{y : (t, y) \in D\}$. If (3.7) and (3.8) hold, then p is a solution of the Feynman-Kac integral equation (3.6), λ^{d+1}-a.e. on D.

Remark 3.4.1 Solutions p investigated in Theorem 3.4.1 are in general not continuous as functions of the space variable. For more restrictive assumptions which provide continuous solutions p on D, we refer to Remark 6.5.1.

Proof. Following Lemma 3.2.5, let K be a compact subset of $\mathbb{R}^d$ such that there exist $s_1 \in (a, b)$ and $\delta_1 > 0$ with

$$\{(s, x) : s \in [s_1 - 2\delta_1, s_1], \, x \in K\} \subset D.$$

Let K_ε be the compact closure of

$$\{y \in \mathbb{R}^d : (v, y) \in \operatorname{supp} j_\varepsilon, \, v \in [s_1 - 2\delta_1, s_1]\}$$

and let

$$\alpha \in C^\infty_{comp}([s_1 - \delta_1, s_1]), \, h \in C^\infty_{comp}(K)$$

play the role of test functions.

In a first step we show that

$$\int ds \, \alpha(s) \int dx \, h(x) \frac{1}{\delta} \int_{s-\delta}^s dv \int g^o(s - \delta, x; v, y) \, dy \, c \, p(v, y) \quad (3.23)$$

$$= \int ds \, \alpha(s) \int dx \, h(x) \, c \, p(s, x) \, j_\varepsilon(s, x) + O(\varepsilon, \tfrac{1}{2}\delta_0) + O(\delta_n, \Delta_n, \varepsilon)$$

where $O(\varepsilon, \tfrac{1}{2}\delta_0) \to 0$ as $\varepsilon \searrow 0$ for $0 < \delta < \tfrac{1}{2}\delta_0 \wedge \delta_n$ and $O(\delta_n, \Delta_n, \varepsilon) \to 0$ as $n \nearrow \infty$. (3.23) is based on a discretization where δ_n and Δ_n denote the maximal sizes of increments in time and space, respectively, which vanish as $n \nearrow \infty$ for any $\varepsilon > 0$. In order to apply the boundary condition (3.25) satisfied by g^o in (3.5), we have to construct a sequence of jointly continuous functions $(p_n^{c,\varepsilon})_{n \in \mathbb{N}}$ on $[s_1 - 2\delta_1, s_1] \times K_\varepsilon$ such that

$$|\sum_{i=1}^{n+1} \delta_n \, \alpha(s_i) \sum_{k \in J_n} \Delta_k^{(n)} \, h(x_k) \frac{1}{\delta_n} \int_{s_i - \delta_n}^{s_i} dv \int dy \, g^o(s_i - \delta_n, x_k; v, y)$$

$$\cdot [c \, p(v, y) - p_n^{c,\varepsilon}(v, y)] \, j_\varepsilon(v, y) | \leq O_1(\delta_n, \Delta_n, \varepsilon) \to 0$$

$$|\sum_{i=1}^{n+1} \delta_n \, \alpha(s_i) \sum_{k \in J_n} \Delta_k^{(n)} \, h(x_k) \, p_n^{c,\varepsilon}(s_i, x_k) \, j_\varepsilon(s_i, x_k)$$

$$- \int ds \, \alpha(s) \int dx \, h(x) \, c(s, x) \, p(s, x) \, j_\varepsilon(s, x) | \leq O_3(\delta_n, \Delta_n, \varepsilon) \to 0$$

as $n \nearrow \infty$ for any $\varepsilon > 0$. The discretization consists of $\delta_n = \tfrac{1}{n}\delta_1$, $s_i = s_1 - (i - 1)\delta_n$, $i = 1, \ldots, n + 1$, and a finite number of disjoint measurable subsets $(V_k^{(n)})_{k \in J_n}$, $J_n \subset \mathbb{N}$, of D such that

$$K = \bigcup_{k \in J_n} V_k^{(n)}$$

with $x_k \in V_k^{(n)}$, $\Delta_k^{(n)} = \lambda^d(V_k^{(n)})$ and $\Delta_n = \max_{k \in J_n} \Delta_k^{(n)} \searrow 0$ as $n \nearrow \infty$.

When cp is approximated by $(p_n^{c,\varepsilon})_{n\in\mathbb{N}}$, the involved weight functions $g^o(s_i - \delta_n, x_k; v, y)$ become singular at $(s_i - \delta_n, x_k) \in [s_1 - 2\delta_1, s_1] \times K$. Following Lemma 3.2.5, we define the family of measures

$$\mu_{s,x,\delta,\varepsilon}(dv\,dy) = \frac{1}{\delta}\, g^o(s - \delta, x; v, y)\, 1_{(s-\delta,s]\times K_\varepsilon}(v, y)\, dv\, dy$$

where

$$cp \in L^1(\mu_{s,x,\delta,\varepsilon})$$

for $(s, x) \in D$ with $s \in [s_1 - \delta_1, s_1]$ and $0 < \delta \le \delta_1$ because of (3.8). Applied to every cell of the discretization, Lemma 3.2.3 and Lemma 3.2.5 yield

$$cp \in L^1(\mu_{\delta_n,\Delta_n,\varepsilon})$$

for

$$\mu_{\delta_n,\Delta_n,\varepsilon} = \sum_{i=1}^{n+1} \sum_{k\in J_n} \mu_{s_i,x_k,\delta_n,\varepsilon} + 1_{[s_1-2\delta_1,s_1]\times K_\varepsilon}(v, y)\, dv\, dy$$

and $n \in \mathbb{N}$. Hence by standard approximation arguments there exists a sequence $(p_{\delta_n,\Delta_n,m}^{c,\varepsilon})_{m\in\mathbb{N}}$ of jointly continuous functions on $[s_1 - 2\delta_1, s_1] \times K_\varepsilon$ such that

$$L^1(\mu_{\delta_n,\Delta_n,\varepsilon}) - \lim_{m\to\infty} p_{\delta_n,\Delta_n,m}^{c,\varepsilon} = cp.$$

Thus the diagonal procedure

$$(s_i(\delta_n), x_k(\Delta_n), p_{\delta_n,\Delta_n,n}^{c,\varepsilon} = p_n^{c,\varepsilon})_{n\in\mathbb{N}}$$

provides

$$L^1(\mu_{\delta_n,\Delta_n,\varepsilon}) - \lim_{n\to\infty} p_n^{c,\varepsilon} = cp.$$

The approximating sequence $p_n^{c,\varepsilon}\, j_\varepsilon$, $n \in \mathbb{N}$, satisfies

$$\sum_{i=1}^{n+1} \delta_n\, \alpha(s_i) \sum_{k\in J_n} \Delta_k^{(n)}\, h(x_k)$$

$$\cdot\, |\, \frac{1}{\delta_n} \int_{s_i-\delta_n}^{s_i} dv \int g^o(s_i - \delta_n, x_k; v, y)\, dy\, p_n^{c,\varepsilon}(v, y)\, j_\varepsilon(v, y)$$

$$- p_n^{c,\varepsilon}(s_i, x_k)\, j_\varepsilon(s_i, x_k)\, | \le O_2(\delta_n, \Delta_n, \varepsilon) \to 0 \qquad (3.24)$$

as $n \nearrow \infty$ for any $\varepsilon > 0$. In fact, following Dynkin (1965) vol. 2, appendix § 6, the boundary condition satisfied by the fundamental solution $g(s, x; t, y)$ of (3.1) means that

$$\lim_{s \nearrow t} \int_{\mathbb{R}^d} g^o(s, x; t, y) \, dy \, j(t, y) = j(t, x), \quad (t, x) \in D, \ s \in (a, t) \qquad (3.25)$$

for bounded and continuous functions j on D where the convergence in x is uniform on compact subsets. Hence there exists $\delta_{0, \varepsilon, n} > 0$ such that

$$\frac{1}{\delta} \int_{s-\delta}^{s} dv \int g^o(s - \delta, x; v, y) \, dy \, p_n^{c, \varepsilon}(v, y) \, j_\varepsilon(v, y)$$
$$= p_n^{c, \varepsilon}(s, x) \, j_\varepsilon(s, x) + O(s, \delta, x, p_n^{c, \varepsilon})$$

for $0 < \delta \leq \delta_{0, \varepsilon, n} \wedge \delta_1$ where $O(s, \delta, x, p_n^{c, \varepsilon}) \to 0$ as $\delta \searrow 0$, uniformly for $(s, x) \in [s_1 - \delta_1, s_1] \times K$. In case of (3.24) we receive

$$O_2(\delta_n, \Delta_n, \varepsilon) = \sup_{\substack{1 \leq i \leq n+1 \\ k \in J_n}} O(s_i, \delta_n, x_k, p_n^{c, \varepsilon}) \, \delta_1 \mid K \mid \max_{[s_1 - \delta_1, s_1] \times K} \alpha \, h$$

which completes the verification of (3.23).

The second step is an immediate consequence of Lemma 3.2.1 and (3.25) claiming that

$$\lim_{\delta \searrow 0} \frac{1}{\delta} \{ \int dx \, h(x) \, g^o(s - \delta, x; s, y) - h(y) \} \qquad (3.26)$$
$$= \frac{1}{2} \Delta_\alpha^*(s, y) \, h(y) - \nabla(\beta(s, y) \, h(y))$$

where the convergence is uniform in y on bounded subsets of $\mathbb{R}^d$.

The third step provides a proof of the first part of the theorem. We replace s by $s - \delta$ as well as u by s in (3.9) and integrate it w.r.t. $\alpha(s) \, h(x) \, ds \, dx$ where $(s, x) \in [s_1 - \delta_1, s_1] \times K$. An application of steps one and two yields

$$\int ds \, \alpha(s) \int dx \, h(x) \, [\frac{1}{\delta} \{ p(s, x) - p(s - \delta, x) \} + c \, p(s, x)] \qquad (3.27)$$
$$+ \int ds \, \alpha(s) \int dy \, p(s, y) \, \{ \frac{1}{2} \Delta_\alpha^*(s, y) \, h(y) - \nabla(\beta(s, y) \, h(y)) \}$$
$$= O(\varepsilon, \frac{1}{2} \delta_0) + O(\delta_n, \Delta_n, \varepsilon) + O(\delta)$$

where the Fubini type argument is justified by Lemma 3.2.3 and $O(\delta) \to 0$ as $0 < \delta \leq \delta_n \searrow 0$. In the limit as n tends to infinity and ε vanishes, (3.27) provides $\partial p / \partial s$, λ^{d+1}-a.e. on D. An integration by parts in (3.27) leads to

$$\int ds \int dx \, p(s, x) \, \{ -\frac{\partial}{\partial s} + \frac{1}{2} \Delta_\alpha^*(s, x) - \nabla(\beta(s, x) \, . \,) + c(s, x) \} \, \gamma(s, x) = 0$$

for functions $\gamma(s, x) = \alpha(s)\, h(x)$ which are dense in $C_{comp}^{\infty}(D)$. The boundary condition (3.22) is an immediate consequence of (3.6), (3.8) and (3.25).

In order to prove the converse part of the theorem, we choose

$$\gamma(t, y) = \int_a^t ds \int \alpha(s)\, h(x)\, dx\, g^o(s, x; t, y) \qquad (3.28)$$

as a test function where $\alpha \in C((a, b))$, $h \in C(\mathbb{R}^d)$ such that $\alpha\, h \in C_{comp}(D)$. This particular test function γ satisfies

$$-\frac{\partial \gamma}{\partial t} + \frac{1}{2}\triangle_\alpha^* \gamma - \nabla(\beta\, \gamma) + \alpha\, h = 0$$

on D with the initial data $\gamma(a) = 0$. Since $\operatorname{supp}\gamma$ is not compact, the assumption means

$$\int dy\, f(t, y)\, \gamma(t, y) + \int_a^t dv \int dy\, p(v, y)\, \{\, -\alpha(v)\, h(y) + c(v, y)\, \gamma(v, y)\, \} = 0.$$

Under (3.7), (3.8) and with (3.28) as well as $\alpha\, h \in C_{comp}(D)$, an application of Fubini's theorem shows that p satisfies (3.6), λ^{d+1}-a.e. on D. $\Diamond$

Remark 3.4.2 The integrability conditions

$$\int_s^t dv\, P_{(s,x)}[|\, (D_s + \frac{1}{2}\triangle_\alpha + \beta\, \nabla)\, \Phi(v, X_v)\, |; v < \tau_s] < \infty \qquad (3.29)$$

and

$$P_{(s,x)}[\Phi(t, X_t); t < \tau_s] < \infty \qquad (3.30)$$

for $(s, x) \in D(\Phi)$, $a \leq s \leq t$, $t \in (a, b]$ fixed, are equivalent to (3.8) and (3.10) in case that p, D and c in Section 3.2 are replaced by a Schrödinger multiplier Φ with state space $D(\Phi)$ in Definition 2.2.2 and the reference potential c given in (2.22).

Corollary 3.4.1 *A Schrödinger multiplier Φ in Definition 2.2.2 which satisfies (3.29) and (3.30) is a solution of the Feynman-Kac integral equation (3.6), if c is the reference potential in (2.22) and $\Phi(t, \cdot)$ appears in place of $f(t, \cdot)$ for fixed $t \in (a, b)$. Moreover, the derivative $\partial\Phi/\partial s$ exists λ^{d+1}-a.e. on $D(\Phi)$.*

Proof. These are consequences of Theorem 3.4.1 and Remark 3.4.2 where we notice that (3.7) can be replaced by (3.10) since Φ is non-negative. $\Diamond$

Corollary 3.4.2 *If a Schrödinger multiplier Φ in Definition 2.2.2 and the reference potential c in (2.22) satisfy*

$$P_{(s,x)}[\exp\{\int_s^t c(u, X_u)\, du\}\, \Phi(t, X_t); t < \tau_s] < \infty \qquad (3.31)$$

and

$$\int_s^t dv\, P_{(s,x)}[\exp\{\int_s^v c(u, X_u)\, du\}\, c^+(v, X_v)\, \Phi(v, X_v); v < \tau_s] < \infty \qquad (3.32)$$

or (3.30) and

$$\int_s^t dv\, P_{(s,x)}[\exp\{\int_s^v |c(u, X_u)|\, du\}\, |c\, \Phi(v, X_v)|; v < \tau_s] < \infty \qquad (3.33)$$

for $(s, x) \in D$, $a < s \le t \le b$, then (2.41), i.e.,

$$P_{(s,x)}[N_s^t] = 1 \quad for \ a < s \le t \le b$$

holds λ^d-a.e. on $\{x \in \mathbb{R}^d : (s, x) \in D(\Phi)\}$ where $(N_s^t)_{s \le t \le b}$ is defined in (2.31).

Remark 3.4.3 There is not an a priori better choice among the two pairs of integrability assumptions (3.31), (3.32) and (3.30), (3.33). The preference depends on the behavior of the reference potential c in (2.22) which is singular in general. (3.33) might usually be the stronger restriction than (3.32) where the negative part c^- supports the integral to be finite.

Proof. Let us first apply Proposition 3.3.1 to Φ in Definition 2.2.2 and c in (2.22). Assumption (3.19) in Proposition 3.3.1 follows from (3.32). Assumption (3.18) claims the finiteness of $\tilde{p}(s, x; t, \Phi(t, .))$ in (3.11). In fact, (2.38) provided by Fatou's lemma in the proof of Corollary 2.3.1 can be written as

$$\tilde{p}(s, x; t, \Phi(t, .)) \le \Phi(s, x)$$

where $\Phi(s, x)$ is finite for $(s, x) \in D(\Phi)$. Moreover, the reference potential c in (2.22) is defined such that Φ is a weak solution of (3.21). The boundary condition at time $t \in (a, b]$ required in Proposition 3.3.1 is satisfied by Φ since it is assumed to be continuous on $D(\Phi)$. Consequently, Φ is a solution of the 'killed' integral equation (3.17).

Second we notice that (3.11) with $\Phi(t, .)$ in place of $f(t, .)$, this means $\tilde{p}(s, x; t, \Phi(t, .))$, is nothing but $p^o(s, x)$ with $\Phi(t, .)$ in place of $p(t, .)$ which we met in the proof of Proposition 3.3.1 as a solution of (3.17). Since Φ satisfies (3.18) and (3.19) as just has been found above, we conclude from $p^o(s,x)$ that $\tilde{p}(s,x;t,\Phi(t, \, . \,))$ is a solution of the 'killed' integral equation (3.17). Thus the uniqueness of solutions of (3.17) stated in Proposition 3.3.1 yields $\tilde{p}(s,x;t,\Phi(t, \, . \,)) = \Phi(s,x)$ for $a \leq s \leq t$ and λ^d-a.a. $x \in I\!\!R^d$ with $(s,x) \in D(\Phi)$ which corresponds to (2.41) because of (2.31).

There is an alternative way of conclusion. Following Remark 3.4.2, Φ satisfies (3.8) and (3.10). Hence they are also true for $\tilde{p}(s,x;t,\Phi(t, \, . \,))$ because of (2.38) in the proof of Corollary 2.3.1. Since $\tilde{p}(s,x;t,\Phi(t, \, . \,))$ is (3.11) with $\Phi(t, \, . \,)$ in place of $f(t, \, . \,)$, it is a solution of the integral equation (3.6) according to Proposition 3.2.1. As a consequence of Theorem 3.4.1 on the equivalence of equations, $\tilde{p}(s,x;t,\Phi(t, \, . \,))$ is a weak solution of (3.21). By assumption and because of (2.38), Φ and $\tilde{p}(s,x;t,\Phi(t, \, . \,))$ satisfy (3.18) and (3.19). Hence Proposition 3.3.1 claims that Φ and $\tilde{p}(s,x;t,\Phi(t, \, . \,))$ are solutions of the 'killed' integral equation (3.17) and (2.41) follows from the uniqueness of solutions.

Finally a more straightforward proof of Corollary 3.4.2 can be given in case of (3.33) which corresponds to (3.13) with Φ and c in (2.22). We notice that (3.33) implies (3.29) which is (3.8). Moreover, $\tilde{p}(s,x;t,\Phi(t, \, . \,))$ which is (3.11) satisfies (3.8) and (3.10) because of Remark 3.4.2 and (2.38). Following Proposition 3.2.1, $\tilde{p}(s,x;t,\Phi(t, \, . \,))$ is a solution of (3.6) with $\Phi(t, \, . \,)$ in place of $f(t, \, . \,)$. By definition of c in (2.22), Φ in Definition 2.2.2 is a weak solution of (3.21) which satisfies (3.22) with $\Phi(t, \, . \,)$ in place of $f(t, \, . \,)$. Hence Theorem 3.4.1 on the equivalence of equations claims that Φ satisfies the same integral equation as $\tilde{p}(s,x;t,\Phi(t, \, . \,))$ does. Consequently, Φ and $\tilde{p}(s,x;t,\Phi(t, \, . \,))$ coincide λ^{d+1}-a.e. on $D(\Phi)$ because of (3.13). $\Diamond$

Chapter 4

Itô's Formula for Non-Smooth Functions

Itô's formula, the chain rule in stochastic calculus, is going to be deduced in case of continuous functions possessing first and second order derivatives only in the sense of distributions. They are evaluated in Section 4.2 along the space-time Brownian motion and in Section 4.3 along non-degenerate locally Hölder-continuous space-time semimartingales which are local diffeomorphisms of a spatial parameter. This kind of results will be obtained by means of forward local $C^{1,\varepsilon}$-semimartingale flows of C^1-diffeomorphisms. First we derive a change of variable formula and show the existence of all moments of the involved Jacobian. Then our version of Itô's formula can be established as a consequence of mollifier properties.

4.1 Meaning and generalization

In the stochastic calculus, Itô's formula plays the role of the common chain rule in analysis. This chapter deals with a version of Itô's formula for continuous functions possessing first and second order derivatives only in the sense of distributions. They are going to take their values along *non-degenerate locally Hölder-continuous space-time semimartingales* which are *local diffeomorphisms* of a spatial parameter. Our investigations will be based on the theory of stochastic flows developed in Kunita (1990). It provides a one-to-one correspondence between stochastic flows and (Itô's) stochastic differential equations if they are formulated in terms of continuous C-valued semimartingales where C denotes a space of continuous functions.

Let us introduce Itô's formula in the most familiar situation. We consider a probability space $(\Omega, \mathcal{F}, (\mathcal{F}_t)_{0 \leq t \leq T}, P)$ endowed with the 1-dimensional Brownian motion $(B_t)_{0 \leq t \leq T}$ where $0 < T < \infty$. Following Itô (1046, 1051, 1061), *Itô's stochastic differential equation*

$$\xi_t(x) = x + \int_0^t \sigma(\xi_s(x)) \, dB_s + \int_0^t \beta(\xi_s(x)) \, ds \tag{4.1}$$

has a unique solution $\xi_t(x)$, $0 \leq t \leq T$, for any $x \in \mathbb{R}$, if the coefficients σ and β are Lipschitz-continuous functions on $\mathbb{R}$ of linear growth. Moreover, ξ_t is a continuous function of the spatial parameter $x \in \mathbb{R}$, P-a.s. for any $t \in [0, T]$. In order to apply Itô's formula to a function F on $\mathbb{R}$ evaluated along the solution ξ_t, $0 \leq t \leq T$, of (4.1), it conventionally has to be assumed that $F \in C^2(\mathbb{R})$. Then

$$F(\xi_t) - F(x) \tag{4.2}$$

$$= \int_0^t F'(\xi_s) \, \sigma(\xi_s) \, dB_s + \int_0^t F'(\xi_s) \, \beta(\xi_s) \, ds + \frac{1}{2} \int_0^t F''(\xi_s) \, \sigma^2(\xi_s) \, ds$$

for $0 \leq t \leq T$, P-a.s. for any $x \in \mathbb{R}$. If the first and second order derivatives $DF \in L^2(\lambda, \mathbb{R})$ and $D^2 F \in L^1(\lambda, \mathbb{R})$, λ the Lebesque measure, of a continuous function F are given in the sense of distributions, i.e.,

$$\int_{\mathbb{R}} DF(x) \, \gamma(x) \, dx = - \int_{\mathbb{R}} F(x) \, \gamma'(x) \, dx \tag{4.3}$$

is satisfied by test functions γ which are smooth and have compact support, then our version of Itô's formula holds analogously to (4.2) for non-degenerate processes $(\xi_t)_{0 \leq t \leq T}$, but only for $\lambda \otimes P$-almost all $(x, \omega) \in \mathbb{R} \times \Omega$.

The proof of our version of Itô's formula is an approximation procedure in terms of a *mollifier sequence* $(J_\varepsilon)_{\varepsilon > 0}$, i.e., $J_\varepsilon F \to F$ uniformly on compact sets, $(J_\varepsilon F)' \to DF$ in $L^2_{loc}(\lambda, \mathbb{R})$ and $(J_\varepsilon F)'' \to D^2 F$ in $L^1_{loc}(\lambda, \mathbb{R})$ as $\varepsilon \searrow 0$. The key to make use of these integrability properties is a change of variable formula: For any $F \in L^1(\lambda, \mathbb{R})$

$$\int_{\mathbb{R}} dx \int_0^T F(\xi_{0,t}(x)) \, dt = \int_0^T dt \int_{\mathbb{R}} dz \, F(z) \det \partial_z \xi_{0,t}^{-1}(z) \tag{4.4}$$

P-a.s. and it is in $L^1(P)$ where $\partial_z \xi_{0,t}^{-1}(z)$ denotes the *Jacobian* of the inverse
of the flow $\xi_{s,t}$, $0 \leq s \leq t \leq T$. We notice that this formula 'decomposes' the
function F and the process ξ_t, i.e., its associated flow $\xi_{s,t}$. For the definition
of stochastic flows we refer to Section 4.3. There is a strong analogy to flows
obtained as solutions of a systems of ordinary differential equations (dynamical systems). In the probabilistic case however, the transport takes place along
solutions of a stochastic differential equations, i.e., along paths which have
non-vanishing *quadratic variation* and are consequently nowhere differentiable.
Nevertheless, the theory of stochastic flows reveals that under smoothness assumptions on the coefficients σ and β, the solution ξ_t, $0 \leq t \leq T$, of (4.1)
depends smoothly on the initial value $x \in I\!\!R$, i.e., the spatial parameter. The
composition $F(\xi_{0,t}(x))$, $0 \leq t \leq T$, is well-defined as a stochastic process if ξ_t,
$0 \leq t \leq T$, is non-degenerate which is satisfied in case of än elliptic σ. The
change of variable formula (4.4) is valid if σ as well as β are bounded and their
derivatives σ', β' are bounded and Hölder-continuous. Due to the flow property,
the Jacobian of the inverse of a stochastic flow is the same as the inverse of
the Jacobian of the stochastic flow. Since the Jacobian of a stochastic flow is
a solution of a system of Itô's stochastic differential equations, all its moments
exist and are uniformly bounded on compact sets of starting points.

4.2   Driving Brownian motion

Let $(\Omega, \mathcal{F}, (\mathcal{F}_t)_{a \leq t \leq b}, P)$ be a probability space endowed with the d-dimensional
Brownian motion $(B_t)_{a \leq t \leq b}$, where $-\infty < a < b < \infty$. Following Kunita
(1990)'s §4.6

$$B_{s,t}(x) = x + B_t - B_s \qquad (4.5)$$

is an elementary example of a continuous forward $C^1([a,b], I\!\!R^d)$-valued Brownian flow of diffeomorphisms for which we obtain

Lemma 4.2.1 (1st change of variable formula) *Let $F \in L_{loc}^1([a,b] \times I\!\!R^d)$ and
let K be a compact subset of $[a,b] \times I\!\!R^d$. Then*

$$\int_{I\!\!R^d} dx \int_s^{\tau_{(s,x)} \wedge b} F(t, B_{s,t}(x)) \, dt \; = \; \int_s^b dt \int_{I\!\!R^d} dz \, F(t, z) \, 1_{[s, \hat{\tau}_{(s,z)}(t)]}(t) \quad (4.6)$$

P-a.s. and it is in $L^1(P)$ where

$$\tau_{(s,x)} = \inf\{u \in [s,b) : (u, B_{s,u}(x)) \notin K\}$$

for $(s, x) \in [a, b] \times I\!\!R^d$ and

$$\hat{\tau}_{(s,z)}(t) \;=\; \tau_{(s,x)} \quad \text{with} \quad x \;=\; B_{s,t}^{-1}(z)$$

for $t \in [s, b]$, P-a.s.

Proof. Following Kunita (1990)'s §1.2, the stochastic process $F(t, B_{s,t}(x))$, $s \leq t \leq h$, is well-defined due to the absolute continuity of the 1-dimensional marginal distributions of $B_{s,t}(x)$ w.r.t. the Lebesgue measure λ^d. The function F is in $L^1(\{z : (t, z) \in K\})$ for λ-a.a. $t \in [a, b]$ and $\tau_{(s,x)}$ is jointly lower semi-continuous in (s, x) as a consequence of Kunita (1990)'s §4.8. Hence

$$\int_{\{x:(s,x)\in K\}} dx \, F(t, B_{s,t}(x)) \, 1_{[s, \tau_{(s,x)}]}(t)$$

$$= \int_{B_{s,t}(\{x:(s,x)\in K\})} dz \, F(t, z) \, 1_{[s, \hat{\tau}_{(s,z)}(t)]}(t) \, \det \partial_z B_{s,t}^{-1}(z)$$

for λ-a.a. $t \in [s, b]$ and P-a.s., where $\partial_z B_{s,t}^{-1}(z)$ denotes the Jacobian of the inverse of $B_{s,t}(x)$ which equals the identity in case of (4.5). Thus

$$\int_s^b dt \int_{I\!\!R^d} dx \mid F(t, B_{s,t}(x)) \mid 1_{[s,\tau_{(s,x)}]}(t) \leq \int_K \int \mid F(t, z) \mid dt \, dz < \infty$$

P-a.s. and Fubini's theorem completes the proof. $\Diamond$

Let D be an open subset of $(a, b) \times I\!\!R^d$. Along the lines of Friedman (1964)'s chap. X, we define generalized derivatives of extended real-valued measurable functions F on D according to (4.3) w.r.t. the space of test functions $C_{comp}^{\infty}(D) = \{\gamma : \gamma$ has a compact support contained in D and its derivatives of any order exist and are continuous$\}$.

Theorem 4.2.1 *Let F be any extended real-valued measurable function on D which possesses $D_s F \in L_{loc}^1(D)$, $D_i F \in L_{loc}^2(D)$ and $D_i^2 F \in L_{loc}^1(D)$ where $D_s = \partial_s$ and $D_i^{\nu} = \partial_{x_i}^{\nu}$, $\nu = 1, 2$, $1 \leq i \leq d$, are to be understood in the sense of distributions w.r.t. $C_{comp}^{\infty}(D)$. It is assumed that $D(F) = \{(s, x) \in D : F$ is jointly continuous at (s, x) and $\mid D_s F(s, x) \mid < \infty$ as well as $\mid D_i^{\nu} F(s, x) \mid < \infty$ for $\nu = 1, 2$, $1 \leq i \leq d\}$ is an open set. Then, for any initial time $s \in (a, b)$ and λ^d-a.e. $x \in I\!\!R^d$,*

$$F(t \wedge \tau_{(s,x)}, B_{s,t \wedge \tau_{(s,x)}}(x)) - F(s, x) = \qquad\qquad (4.7)$$

$$= \int_s^{t\wedge\tau_{(s,x)}} D_sF(u, B_{s,u}(x))\, du + \sum_{i=1}^d \int_s^{t\wedge\tau_{(s,x)}} D_iF(u, B_{s,u}(x))\, dB_u^i$$

$$+ \frac{1}{2}\sum_{i=1}^d \int_s^{t\wedge\tau_{(s,x)}} D_i^2F(u, B_{s,u}(x))\, du$$

for $t \in [s, b)$, *P-a.s., with* $\tau_{(s,x)} = \inf\{u \in [s,b) : (u, B_{s,u}(x)) \notin K(F)\}$ *where* $K(F)$ *is any compact subset of* $D(F)$.

Proof. The mollifier theory in Friedman (1964)'s chap. X provides for each $K(F)$ a sequence $(J_\varepsilon)_{\varepsilon>0}$ of smoothing operators such that

$$
\begin{aligned}
J_\varepsilon F &\to F \text{ uniformly on } K(F)\\
D_sJ_\varepsilon F = J_\varepsilon D_s F &\to D_s F \text{ in } L^1(K(F))\\
D_iJ_\varepsilon F = J_\varepsilon D_i F &\to D_i F \text{ in } L^2(K(F))\\
D_i^2 J_\varepsilon F = J_\varepsilon D_i^2 F &\to D_i^2 F \text{ in } L^1(K(F))
\end{aligned}
$$

as $\varepsilon \searrow 0$ for $1 \le i, j \le d$. Hence the classical Itô formula claims that

$$J_\varepsilon F(t \wedge \tau_{(s,x)}, B_{s,t\wedge\tau_{(s,x)}}(x)) - J_\varepsilon F(s, x) \tag{4.8}$$

$$= \int_s^{t\wedge\tau_{(s,x)}} \frac{\partial}{\partial s} J_\varepsilon F(u, B_{s,u}(x))\, du + \sum_{i=1}^d \int_s^{t\wedge\tau_{(s,x)}} \frac{\partial}{\partial x_i} J_\varepsilon F(u, B_{s,u}(x))\, dB_u^i$$

$$+ \frac{1}{2}\sum_{i=1}^d \int_s^{t\wedge\tau_{(s,x)}} \frac{\partial^2}{\partial x_i^2} J_\varepsilon F(u, B_{s,u}(x))\, du$$

for $t \in [s, b)$, *P*-a.s. The investigation for $\varepsilon \searrow 0$ yields in case of the stochastic integral on the right-hand side of (4.8) that

$$\int_{\mathbb{R}^d} dx\, P[\int_s^{t\wedge\tau_{(s,x)}} |D_iF(u, B_{s,u}(x))|^2\, du]$$

$$= P[\int_s^t du \int_{\mathbb{R}^d} dz\, |D_iF(u, z)|^2\, 1_{[s,\hat{\tau}_{(s,z)}(u)]}(u)] < \infty$$

because of the change of variable formula (4.6). Hence

$$L^2(\lambda^d \otimes P) - \lim_{\varepsilon\searrow 0} \int_s^{t\wedge\tau_{(s,x)}} \partial_{x_i} J_\varepsilon F(u, B_{s,u}(x))\, dB_u^i$$

$$= \int_s^{t\wedge\tau_{(s,x)}} D_iF(u, B_{s,u}(x))\, dB_u^i \in L^2(\lambda^d \otimes P)$$

for $t \in [s, b)$ is a consequence of the definition of stochastic integrals in Ikeda-Watanabe (1989)'s chap. II, definition 1.5. By means of Chebyshev's inequality and the Borel-Cantelli lemma, a subsequence $(\varepsilon_{1,n})_{n \in \mathbb{N}}$ can be found along which this convergence takes place $\lambda^d \otimes P$-a.s. for $1 \leq i \leq d$.

The $L^1(\lambda^d \otimes P)$-convergence of the two bounded variation members of the right-hand side of (4.8) follows from the change of variable formula (4.6), too. The first term satisfies

$$\int dx \, P[\int_s^{t \wedge \tau(s,x)} \mid \partial_s J_\varepsilon F(u, B_{s,u}(x)) - D_s F(u, B_{s,u}(x)) \mid du]$$
$$\leq \quad \parallel \partial_s J_\varepsilon F - D_s F \parallel_{L^1(K(F))} \to 0.$$

for $t \in [s, b)$ as $\varepsilon \searrow 0$; the second one is treated analogously. Hence a subsequence $(\varepsilon_{2,n})_{n \in \mathbb{N}} \subset (\varepsilon_{1,n})_{n \in \mathbb{N}}$ exists such that the right-hand side of (4.8) converges to the right-hand side of (4.7) for $t \in [s, b)$ as $n \nearrow \infty$, $\lambda^d \otimes P$-a.s., i.e., for $\lambda^d \otimes P$-almost all $(x, \omega) \in \mathbb{R}^d \times \Omega$. Equation (4.8) is true for $t \in [s, b)$ and all $x \in \mathbb{R}^d$, P-a.s., and its left-hand side tends to the left-hand side of (4.7) for $t \in [s, b)$ as $n \nearrow \infty$, P-a.s. Thus, (4.7) holds for $t \in [s, b)$, $\lambda^d \otimes P$-a.s. $\Diamond$

4.3 Driving flows of diffeomorphisms

Let $(\Omega, \mathcal{F}, (\mathcal{F}_t)_{a \leq t \leq b}, P)$ be a probability space, where $-\infty < a < b < \infty$, and let

$$I(x, t) \; = \; M(x, t) + B(x, t) \tag{4.9}$$

$a \leq t \leq b$, $x \in \mathbb{R}^d$, be a continuous, d-dimensional, non-degenerate C-semimartingale. This means that $I(\, . \, , t)(\omega)$ is continuous on $\mathbb{R}^d$ for any $t \in [a, b]$ and P-a.s. Moreover, $M(x, t)$, $a \leq t \leq b$, has to be a continuous localmartingale and $B(x, t)$, $a \leq t \leq b$, has to be a continuous process of bounded variation for any value of the spatial parameter $x \in \mathbb{R}^d$. Following Kunita (1990)'s §3, there exist a continuous increasing process A_t and a family of predictable processes $\alpha(x, y, t)$, $x, y \in \mathbb{R}^d$, such that the joint quadratic variation of $I(x, t)$ can be expressed as

$$< I(x, t), I(y, t) > \; = \; < M(x, t), M(y, t) > \; = \; \int_a^t \alpha(x, y, u) \, dA_u \, , \quad P\text{-a.s.}$$

Hence we assume that the process $B(x, t)$, $a \le t \le b$, can be given as

$$B(x, t) = \int_a^t \beta(x, u) \, dA_u \,, \quad P\text{-a.s.}$$

by a family of predictable processes $\beta(x, t)$, $x \in \mathbb{R}^d$.

Assumption $(\mathcal{A})$: $\alpha(x, y, t)$, $\beta(x, t)$ and $\frac{\partial^2}{\partial x_i \partial y_j} \alpha(x, y, t)$, $\frac{\partial}{\partial x_i} \beta(x, t)$, $1 \le i, j \le d$, are uniformly in $\omega \in \Omega$ bounded on compact subsets of $(a, b) \times \mathbb{R}^d$ where $\alpha(x, y, t)$ satisfies a uniform ellipticity condition. Moreover, $\frac{\partial^2}{\partial x_i \partial y_j} \alpha(x, y, t)$ and $\frac{\partial}{\partial x_i} \beta(x, t)$, $1 \le i, j \le d$, are locally δ-Hölder continuous in (x, y) and x, respectively, for a $\delta \in (0, 1]$ and for random Hölder coefficients in $L^1(A_t)$.

Under $(\mathcal{A})$, Kunita (1990)'s §4.7 claims that Itô's stochastic differential equation

$$\xi_{s,t}(x) = x + \int_s^t I(\xi_{s,u}(x), du) \tag{4.10}$$

defines on every compact set $K \subset (a, b) \times \mathbb{R}^d$ a unique P-a.s. continuous forward $C^{1,\varepsilon}$-semimartingale flow of local C^1-diffeomorphisms $\xi_{s,t}(x)$, $(s, x) \in K$, $s \le t \le \tau_{(s,x)}^K$, for $\varepsilon < \delta$, where $\tau_{(s,x)}^K = \inf\{t \ge s : (t, \xi_{s,t}(x)) \notin K\} \le b$ for $(s, x) \in (a, b) \times \mathbb{R}^d$. This means:

i) $\xi_{s,t}(x)$, $s \le t \le \tau_{(s,x)}^K$, is a continuous semimartingale.

ii) $\xi_{s,t} : \{x : (s, x) \in \text{int}(K), \, t < \tau_{(s,x)}^K\} \to \mathbb{R}^d$ is once ε-Hölder continuously differentiable for any $t \in (s, b]$.

iii) $\xi_{s,t}(x) = \xi_{u,t}(\xi_{s,u}(x))$ on $\{x : (s, x) \in K, \, t \le \tau_{(s,x)}^K\}$ for $s < u < t$ which is called (local) flow property.

iv) $\xi_{s,s}(x)$ is the identity for all $s \in (a, b)$ with $(s, x) \in K$.

v) $\xi_{s,t} : \{x : (s, x) \in \text{int}(K), \, t < \tau_{(s,x)}^K\}$
    $\to \{\xi_{s,t}(x) : (s, x) \in \text{int}(K), \, t < \tau_{(s,x)}^K\}$
is a C^1-diffeomorphism for any $t \in (s, b]$.

Evidently, all properties i) - v) depending on $\omega \in \Omega$ hold P-a.s.

Lemma 4.3.1 (2nd change of variable formula) *Let $(\mathcal{A})$ be assumed and let $F \in L^1_{loc}((a, b) \times \mathbb{R}^d)$. Then, for any compact $K \subset (a, b) \times \mathbb{R}^d$,*

$$\int_{\mathbb{R}^d} dx \int_s^{\tau_{(s,x)}^K} F(t, \xi_{s,t}(x)) \, dt \tag{4.11}$$

$$= \int_s^b dt \int_{\mathbb{R}^d} dz \, F(t, z) \, 1_{[s, \hat{\tau}_{(s,z)}^K]}(t) \, \det \partial_z \xi_{s,t}^{-1}(z)$$

P-a.s. and it is in $L^1(P)$ where

$$\hat{\tau}^K_{(s,z)} = \tau^K_{(s,x)} \quad with \quad x = \xi^{-1}_{s,t}(z)$$

for $t \in [s, b]$, P-a.s., and $\partial_z \xi^{-1}_{s,t}(z)$ denotes the Jacobian of the inverse of the flow $\xi_{s,t}(x)$.

Proof. The stochastic process $F(t, \xi_{s,t}(x))$, $s \le t < b$, is well-defined in the sense of Kunita (1990)'s §1.2 since the 1-dimensional marginal distributions of any 1-point motion of $\xi_{s,t}(x)$, $s \le t < b$, are absolutely continuous w.r.t. λ^d. Let a compact set $K \subset (a, b) \times \mathbb{R}^d$ be fixed. Then $F \in L^1(\{z : (t, z) \in K\})$ for λ-a.a. $t \in (a, b)$ and the local C^1-diffeomorphic property of the flow $\xi_{s,t}$ yields

$$\int_{\{x:(s,x)\in K\}} dx\, F(t, \xi_{s,t}(x))\, 1_{[s,\tau^K_{(s,x)}]}(t) \tag{4.12}$$

$$= \int_{\xi_{s,t\wedge\tau^K_{(s,x)}}(\{x:(s,x)\in K\})} dz\, F(t, z)\, 1_{[s,\hat{\tau}^K_{(s,z)}]}(t)\, \det \partial_z \xi^{-1}_{s,t}(z)$$

for λ-a.a. $t \in (a, b)$ and P-a.s., where $\hat{\tau}^K_{(s,z)}$ in terms of $\xi_{s,t}(x)$ is well-defined. Since $\xi_{s,t}(x)$ is differentiable in x, the flow property provides

$$\partial_z \xi^{-1}_{s,t}(z)\, |_{z=\xi_{s,t}(x)} = (\partial_x \xi_{s,t}(x))^{-1} \tag{4.13}$$

on $\{t < \tau^K_{(s,x)}\}$, P-a.s. We notice that the compact set

$$G_s = \{\, (t \wedge \tau^K_{(s,x)}, \xi_{s,t\wedge\tau^K_{(s,x)}}(x)) : (s, x) \in K\,,\ s \le t \le b\}$$

is P-a.s. homeomorphic to

$$H_s = \{\, (t, x) : (s, x) \in K\,,\ s \le t \le \tau^K_{(s,x)}\,\}.$$

Hence

$$\max_{(t,z)\in G_s} |\det \partial_z \xi^{-1}_{s,t}(z)| = \max_{(t,x)\in H_s} |\det \partial_x \xi_{s,t}(x)|^{-1} < \infty\,, \quad P\text{-a.s.}$$

due to the space-time continuity of the Jacobian of the stochastic flow $\xi_{s,t}(x)$. From (4.12) we receive

$$\int_s^b dt \int_{\mathbb{R}^d} dx\, |\, F(t, \xi_{s,t}(x))\, |\, 1_{[s,\tau^K_{(s,x)}]}(t)$$

$$\le \max_{(t,x)\in H_s} |\det \partial_x \xi_{s,t}(x)|^{-1} \int_K \int |\, F(t, z)\, |\, dt\, dz < \infty\,, \quad P\text{-a.s.}$$

An application of Fubini's theorem w.r.t. P completes the proof provided that

$$\sup_{(t,z)\in K} P[1_{[s,\hat{\tau}^K_{(s,z)}]}(t) \mid \det \partial_z \xi^{-1}_{s,t}(z) \mid] < \infty.$$

Because of (4.13) it is sufficient to show that

$$\sup_{\{(t,x):(s,x)\in K,\ s\leq t\leq b\}} P[\parallel (\partial_x \xi_{s,t\wedge\tau^K_{(s,x)}}(x))^{-1} \parallel^r] < \infty \quad \text{for } 1 \leq r < \infty. \quad (4.14)$$

Following Kunita (1990)'s theorem 3.3.3, $(\partial_x \xi_{s,t}(x))^{-1} = U(x,t)$ satisfies on K

$$
U_{ij}(x,t) = \delta_{ij} - \int_s^t \sum_{l=1}^d U_{il}(x,u) \frac{\partial}{\partial x_j} I^l(\xi_{s,u}(x), du)
$$

$$
+ \int_s^t \sum_{l=1}^d U_{il}(x,u) \sum_{m=1}^d \frac{\partial^2}{\partial x_j \partial y_m} a^{ml}(\xi_{s,u}(x), \xi_{s,u}(x), u) \, dA_u
$$

P-a.s. for $1 \leq i,j \leq d$ which forms together with (4.10) a system of d^2+d Itô's stochastic differential equations. Under $(\mathcal{A})$, Kunita (1990)'s theorem 3.4.6 can be adapted to the local situation of this system. As a consequence, its solutions have finite moments of any order which depend continuously on the starting points. Hence (4.14) follows. $\Diamond$

Let us now deduce a version of Itô's formula for semimartingale flows of diffeomorphisms as a consequence of the 2nd change of variable formula (4.11).

Theorem 4.3.1 *Under $(\mathcal{A})$, let $I(x,t)$ be the continuous C-semimartingale in (4.9) and let $\xi_{s,t}(x)$ be the stochastic flow which is the solution of Itô's stochastic differential equation (4.10) driven by $I(x,t)$. Moreover, let F be any extended real-valued measurable function on D which possesses $D_s F \in L^1_{loc}(D)$, $D_i F \in L^2_{loc}(D)$ and $D_i D_j F \in L^1_{loc}(D)$ for $1 \leq i,j \leq d$ where $D_s = \partial_s$ and $D_i = \partial_{x_i}$ are to be understood in the sense of distributions w.r.t. $C^\infty_{comp}(D)$. It is assumed that $D(F) = \{(s,x) \in D : F \text{ is jointly continuous at } (s,x) \text{ and } \mid D_s F(s,x) \mid< \infty, \mid D_i F(s,x) \mid< \infty \text{ as well as } \mid D_i D_j F(s,x) \mid< \infty \text{ for } 1 \leq i,j \leq d\}$ is an open set. Then, for any $s \in (a,b)$,*

$$
F(t \wedge \tau_{(s,x)}, \xi_{s,t\wedge\tau_{(s,x)}}(x)) - F(s,x) \quad (4.15)
$$

$$
= \int_s^{t\wedge\tau_{(s,x)}} D_s F(u, \xi_{s,u}(x)) \, du +
$$

$$+ \quad \sum_{i=1}^{d} \int_{s}^{t \wedge \tau_{(s,x)}} D_i F(u, \xi_{s,u}(x)) \, I^i(\xi_{s,u}(x), du)$$

$$+ \quad \frac{1}{2} \sum_{i,j=1}^{d} \int_{s}^{t \wedge \tau_{(s,x)}} D_i D_j F(u, \xi_{s,u}(x)) \, \alpha^{ij}(\xi_{s,u}(x), \xi_{s,u}(x), u) \, dA_u$$

for $t \in [s,b)$, $\lambda^d \otimes P$-a.s. with

$$\tau_{(s,x)} \; = \; \inf\{u \in [s,b) : (u, \xi_{s,u}(x)) \notin K(F)\}$$

where $K(F)$ is any compact subset of $D(F)$.

Corollary 4.3.1 *Itô's formula (4.15) can be integrated w.r.t. any bounded measurable function on $\mathbb{R}^d$ with compact support. Then, the equality holds P-a.s. and the members are in $L^1(P)$.*

Proof. The mollifier theory developed in Friedman (1964)'s chap. X provides for each $K(F)$ a sequence $(J_\varepsilon)_{\varepsilon>0}$ of smoothing operators such that $J_\varepsilon F \to F$ uniformly on $K(F)$, $D_s J_\varepsilon F \to D_s F$ in $L^1(K(F))$, $D_i J_\varepsilon F \to D_i F$ in $L^2(K(F))$ and $D_i D_j J_\varepsilon F \to D_i D_j F$ in $L^1(K(F))$ as $\varepsilon \searrow 0$ for $1 \le i,j \le d$ where J_ε commutes with D_s and D_i, respectively. Following Kunita (1990)'s theorem 4.7.2, Itô's formula for smooth functions evaluated along stochastic flows may be applied locally on $K(F)$. It yields

$$J_\varepsilon F(t \wedge \tau_{(s,x)}, \xi_{s,t \wedge \tau_{(s,x)}}(x)) - J_\varepsilon F(s,x) \qquad (4.16)$$

$$= \int_{s}^{t \wedge \tau_{(s,x)}} \frac{\partial}{\partial s} J_\varepsilon F(u, \xi_{s,u}(x)) \, du$$

$$+ \quad \sum_{i=1}^{d} \int_{s}^{t \wedge \tau_{(s,x)}} \frac{\partial}{\partial x_i} J_\varepsilon F(u, \xi_{s,u}(x)) \, I^i(\xi_{s,u}(x), du)$$

$$+ \quad \frac{1}{2} \sum_{i,j=1}^{d} \int_{s}^{t \wedge \tau_{(s,x)}} \frac{\partial^2}{\partial x_i \, \partial x_j} J_\varepsilon F(u, \xi_{s,u}(x)) \, \alpha^{ij}(\xi_{s,u}(x), \xi_{s,u}(x), u) \, dA_u$$

for $t \in [s,b)$, P-a.s.

The $\mathcal{F}_t$-adapted process A_t, $s \le t \le b$, is continuous and increasing. Hence we can employ the change of time scale in Kunita (1990)'s §3.2. The time shift

$$\theta_s(u) = \sup\{t \in [s,b] : A_t \le u\} \quad \text{for} \ \ u \in [A_s, \infty)$$

is a continuous *stopping time*. It represents the inverse of A_t in the form of

$$A_{\theta_s(u)} = u, \quad \text{if } A_s \leq u < A_b$$
$$A_{\theta_s(u)} = A_b, \quad \text{if } u \geq A_b$$

as well as

$$\theta_s(A_t) = t, \quad \text{if } s \leq t \leq b.$$

In case of

$$\mathcal{F}_u^{\theta_s} = \mathcal{F}_{\theta_s(u)} \quad \text{for } u \in [A_s, \infty)$$

Kunita (1990)'s theorem 3.2.8 claims that if $I(x, t)$, $x \in \mathbb{R}^d$, is a continuous $C^{1,\delta}$-semimartingale adapted to $(\mathcal{F}_t)_{s \leq t \leq b}$, then

$$I^{\theta_s}(x, u) = I(x, \theta_s(u)) \quad \text{for } u \in [A_s, \infty)$$

is again a continuous $C^{1,\delta}$-semimartingale, but now adapted to $(\mathcal{F}_u^{\theta_s})_{A_s \leq u < \infty}$. Moreover, the *local characteristic* $(a(x, y, t), b(x, t), A_t)$ of $I(x, t)$, $x \in \mathbb{R}^d$, $s \leq t \leq b$, is transformed under the time shift θ_s to the local characteristic $(a^{\theta_s}(x, y, t), b^{\theta_s}(x, t), A_t^{\theta_s})$ of $I^{\theta_s}(x, u)$, $A_s \leq u < \infty$, according to

$$a^{\theta_s}(x, y, u) = a(x, y, \theta_s(u))$$
$$b^{\theta_s}(x, u) = b(x, \theta_s(u))$$
$$A_u^{\theta_s} = A_{\theta_s(u)} = u, \quad \text{if } A_s \leq u < A_b$$
$$= A_b, \quad \text{if } u \geq A_b.$$

Let us denote

$$\xi_{A_s, u}^{\theta_s, x} = \xi_{s, \theta_s(u)}^x \quad \text{for } u \in [A_s, \infty)$$

where

$$\xi_{A_s, A_s}^{\theta_s, x} = \xi_{s, \theta_s(A_s)}^x = x.$$

As a consequence of Kunita (1990)'s theorem 3.2.9, Itô's stochastic differential equation (4.10) can be written as

$$\int_{A_s}^u I^{\theta_s}(\xi_{A_s, r}^{\theta_s, x}, dr) = \int_{\theta_s(A_s) = s}^{\theta_s(u)} I(\xi_{s, r}^x, dr) = \xi_{s, \theta_s(u)}^x = \xi_{A_s, u}^{\theta_s, x}$$

for $A_a \leq u < \infty$. Hence Kunita (1990)'s lemma 3.2.2 claims that the local characteristic of $\xi_{s, t}^x$, $s \leq t \leq b$, is

$$(a(\xi_{s, t}^x, \xi_{s, t}^x, t), b(\xi_{s, t}^x, t), A_t).$$

Thus the local characteristic of $\xi_{A_s,u}^{\theta_s,x}$, $A_s \leq u < \infty$, takes the form

$$(a(\xi_{A_s,u}^{\theta_s,x}, \xi_{A_s,u}^{\theta_s,x}, \theta_s(u)), b(\xi_{A_s,u}^{\theta_s,x}, \theta_s(u)), A_{\theta_s(u)} = u).$$

As a conclusion we may assume without loss of generality that $A_t = t$ for $s \leq t \leq b$. The general case follows by speeding up the time scale according to A_t, $s \leq t \leq b$.

Equation (4.16) has to be treated for $\varepsilon \searrow 0$. In case of the stochastic integral on its right-hand side we receive

$$\int_{I\!\!R^d} dx\, P[\int_s^{t \wedge \tau(s,x)} \mid D_i F(u, \xi_{s,u}(x)) \mid^2 \alpha^{ii}(\xi_{s,u}(x), \xi_{s,u}(x), u)\, du]$$

$$\leq \sup_{(u,z) \in K(F)} \alpha^{ii}(z, z, u)$$

$$\cdot P[\int_s^t du \int_{I\!\!R^d} dz \mid D_i F(u, z) \mid^2 1_{[s,\hat{\tau}(s,z)]}(u) \mid \det \partial_z \xi_{s,u}^{-1}(z) \mid] < \infty$$

by means of the change of variable formula (4.11). As a consequence of the definition of stochastic integrals in terms of L^2-isomorphisms,

$$L^2(\lambda^d \otimes P) - \lim_{\varepsilon \searrow 0} \int_s^{t \wedge \tau(s,x)} \frac{\partial}{\partial x_i} J_\varepsilon F(u, \xi_{s,u}(x))\, M^i(\xi_{s,u}(x), du)$$

$$= \int_s^{t \wedge \tau(s,x)} D_i F(u, \xi_{s,u}(x))\, M^i(\xi_{s,u}(x), du) \in L^2(\lambda^d \otimes P)$$

follows for $t \in [s, b)$. Chebyshev's inequality and the Borel-Cantelli lemma yield a subsequence $(\varepsilon_{1,n})_{n \in I\!\!N}$ for which the convergence holds $\lambda^d \otimes P$-almost surely for every $1 \leq i \leq d$. The three bounded variation members of the right-hand side of (4.16) are treated similarly. Their $L^1(\lambda^d \otimes P)$-convergence follows by means of the change of variable formula (4.11). In case of the third member we obtain

$$\int dx\, P[\int_s^{t \wedge \tau(s,x)} \mid \frac{\partial^2}{\partial x_i\, \partial x_j} J_\varepsilon F(u, \xi_{s,u}(x)) - D_i D_j F(u, \xi_{s,u}(x)) \mid$$

$$\cdot \mid \alpha^{ij}(\xi_{s,u}(x), \xi_{s,u}(x), u) \mid\, du]$$

$$\leq \sup_{(u,z) \in K(F)} (\mid \alpha^{ij}(z, z, u) \mid P[1_{[s,\hat{\tau}(s,z)]}(u) \mid \det \partial_z \xi_{s,u}^{-1}(z) \mid])$$

$$\cdot \parallel \frac{\partial^2}{\partial x_i\, \partial x_j} J_\varepsilon F - D_i D_j F \parallel_{L^1(K(F))} \to 0$$

as $\varepsilon \searrow 0$ for $1 \leq i, j \leq d$. With the same method as used to extract $(\varepsilon_{1,n})_{n \in I\!\!N}$, a subsequence $(\varepsilon_{2,n})_{n \in I\!\!N} \subset (\varepsilon_{1,n})_{n \in I\!\!N}$ can be found along which the convergence on the right-hand side of (4.16) for $t \in [s, b)$ takes place $\lambda^d \otimes P$-almost surely. This follows for the left-hand side of (4.16) from the uniform convergence of $J_\varepsilon F$ to F on $K(F)$. Consequently, given (4.16) which holds P-a.s. for all $x \in \mathbb{R}^d$, formula (4.15) is true on $K(F)$, $\lambda^d \otimes P$-a.s. $\diamond$

Remark 4.3.1 Krylov (1980) develops a version of Itô's formula for continuous functions F with generalized derivatives. Given a bounded state space E, he assumes that $F \in L^p(E)$ where p has to be greater or equal to the dimension of E. Moreover, the coefficients of the underlying diffusion process are related to each other. Then Krylov's version of Itô's formula holds for all starting points and it is proved by geometrical arguments as well as substantial results from the theory of parabolic differential equations. Carmona-Nualart (1990) adopt Kunita (1990)'s theory of stochastic flows and derive an approach in terms of so-called stochastic integrators.

Chapter 5

Large Deviations

Schrödinger processes in Definition 2.4.1 are diffusion processes with given dynamics and prescribed marginal distributions at finite initial time a and final time b. Section 1.3 provides a large deviation approach to Schrödinger processes in a discrete setting. This chapter is devoted to a large deviation principle for the set $A_{a,b}$ of probability measures on a path space which have the prescribed pair (μ_a, μ_b) of probability measures as initial and final distribution, respectively. The formulation of a so-called approximate Sanov property of $A_{a,b}$ in Section 5.1 is not straightforward. In fact, its interior $A_{a,b}$ depending on the topology can be empty and empirical distributions taking only discrete values do not belong to the set $A_{a,b}$ in general. Hence an approximation of $A_{a,b}$ by enlarged sets of probability measures is going to be analyzed by means of Csiszar's τ_0-topology.

The rate function of the large deviation principle for the set $A_{a,b}$ turns out to be the minimal relative entropy with respect to a reference Markov process $\overline{P}$ which represents the given dynamics. The probability measure Q which minimizes the rate function over $A_{a,b}$ will be called Csiszar's projection of $\overline{P}$ on $A_{a,b}$. It is in a sense 'most probable' in $A_{a,b}$ and supposed to be the Schrödinger process determined by μ_a, μ_b and the dynamics of $\overline{P}$. An explicit identification will be deduced later in Section 6.2. Large deviations of Schrödinger processes are investigated in Föllmer (1988) for Brownian motions, i.e., with no creation and killing, and in Dawson-Gorostiza-Wakolbinger (1990) for bounded creation and killing. In both papers however, the technical difficulties related to the Sanov property of the set $A_{a,b}$ are not fully discussed.

The n-product $\overline{\mathbf{P}}^{(n,k)}$ of the reference process $\overline{P}$ conditioned by means of the empirical distribution on an approximation of $A_{a,b}$ is established in Section 5.1. Section 5.2 shows that $\overline{\mathbf{P}}^{(n,k)}$ is weakly asymptotically quasi-independent with Csiszar's projection Q as limiting distribution. Although $\overline{\mathbf{P}}^{(n,k)}$ is not Markovian in general, we can construct a Markovian modification which represents a system of interacting diffusion processes. Its weakly asymptotical quasi-independence will be the key to formulate propagation of chaos in entropy in Section 6.3.

5.1   Approximate Sanov property

Let $\Omega = C([a,b], \mathbb{R}^d)$, $-\infty < a < b < \infty$, be the space of continuous paths taking values in $\mathbb{R}^d$, endowed with the Borel σ-field $\sigma(\Omega)$. The state of a path $X \in \Omega$ at time t, $a \leq t \leq b$, is denoted by $X(t) = X_t$. Let $M_1(\Omega)$ be the space of probability measures on Ω endowed with *Csiszar's τ_0-topology* : For $\varepsilon > 0$ and finite $\sigma(\Omega)$-measurable partitions $\mathcal{P}_k(\Omega) = \{\Omega_1 \ldots \Omega_k\}$, $k \in \mathbb{N}$, of Ω, the basic neighborhoods of an element $P \in M_1(\Omega)$ are defined in Csiszar (1984) by

$$U_0(P, \varepsilon, \mathcal{P}_k) \quad = \quad \{R \in M_1(\Omega) : \mid R(\Omega_i) - P(\Omega_i) \mid < \varepsilon, \; i = 1, \ldots, k, \quad (5.1)$$
$$\text{and } R << P \text{ on } \sigma(\mathcal{P}_k)\}$$

where $\sigma(\mathcal{P}_k)$ is the (trivial) σ-field generated by the partition $\mathcal{P}_k$.

For a given pair of probability measures (μ_a, μ_b) on $\mathbb{R}^d$ we consider

$$A_{a,b} = \{P \in M_1(\Omega) : P \circ X_r^{-1} = q_r \quad \text{for} \;\; r = a, b\} \subset M_1(\Omega) \qquad (5.2)$$

which represents the class of continuous stochastic processes with *prescribed marginal distributions* μ_a and μ_b at finite initial time a and final time b, respectively.

Let a fixed Markov process $\overline{P} \in M_1(\Omega)$ play the role of a *reference measure* on Ω. Examples are renormalized diffusion processes with creation and killing as they will occurs in Section 6.2 when dealing with diffusions related to Schrödinger equations. We assume that

$$\exists \, P \in A_{a,b} \; : \; H(P \mid \overline{P}) < \infty \qquad (5.3)$$

where the *relative entropy* $H(P \mid \overline{P})$ of P with respect to $\overline{P}$ is defined as

$$H(P \mid \overline{P}) \;=\; \int \log(\frac{dP}{d\overline{P}})\, dP\,, \quad \text{if } P << \overline{P} \tag{5.4}$$

$$=\; \infty\,, \qquad\qquad \text{otherwise.}$$

In most of the interesting known situations and hence also in our investigations, the measure $\overline{P}$ itself is not an element of the set $A_{a,b}$.

Let $(\Omega^n, \overline{\mathbf{P}})$ be n independent copies of $(\Omega, \overline{P})$ which means that $\overline{\mathbf{P}}$ is the n-product of the probability measure $\overline{P}$. We consider the *empirical distribution* of $\mathbf{X} = (X^{(1)}, \dots, X^{(n)}) \in \Omega^n$ as a probability measure on Ω denoted by

$$L_n(\mathbf{X}) = \frac{1}{n} \sum_{i=1}^{n} \delta_{X^{(i)}}\,. \tag{5.5}$$

In order to give a conditioning we would like the empirical distribution L_n to satisfy the marginal conditions in question. However, $\{L_n \in A_{a,b}\}$ might be empty in general since L_n takes only discrete values. Hence the set $A_{a,b}$ has to be approximated by enlarged sets. Let $\mathcal{P}_k(\mathbb{R}^d) = \{B_1, \dots, B_k\}$, $k \in \mathbb{N}$, be a sequence of finite measurable partitions of $\mathbb{R}^d$ such that

$$\sigma(\mathcal{P}_k(\mathbb{R}^d)) \subset \sigma(\mathcal{P}_{k+1}(\mathbb{R}^d)) \quad \nearrow \sigma(\mathbb{R}^d) \quad \text{as } k \nearrow \infty. \tag{5.6}$$

We define a family of subsets $A(\varepsilon, k)$ of $M_1(\Omega)$ for $\varepsilon > 0$ and $k \in \mathbb{N}$ in terms of the partitions $\mathcal{P}_k(\mathbb{R}^d)$ as

$$A(\varepsilon, k) \;=\; \{P \in M_1(\Omega) :\mid P(X_r \in B_i) - q_r(B_i) \mid \leq \frac{\varepsilon}{2^k},\; i = 1, \dots, k, \tag{5.7}$$

$$\text{and } P \circ X_r^{-1} << \overline{P} \circ X_r^{-1} \text{ on } \sigma(\mathcal{P}_k(\mathbb{R}^d)) \text{ for } r = a, b \,\}.$$

In view of a conditioning it is to ensure that

$$\overline{\mathbf{P}}(L_n \in A(\varepsilon, k)) > 0 \tag{5.8}$$

for $\varepsilon > 0$, $k \in \mathbb{N}$ and $n \in \mathbb{N}$ sufficiently large. Let us fix a partition $\mathcal{P}_k(\mathbb{R}^d) = \{B_i : i = 1, \dots, k\}$, $k \in \mathbb{N}$, in (5.6). Since $L_n(\mathbf{X}) \circ X_r^{-1}$ for $\mathbf{X} \in \Omega^n$ is concentrated on $X_r^{(j)} \in \mathbb{R}^d$, $j = 1, \dots, n$, $\{L_n \in A(\varepsilon, k)\}$ is the set of those $\mathbf{X} \in \Omega^n$ for which

$$\mid \frac{\sharp_r(B_i)}{n} - q_r(B_i) \mid \leq \frac{\varepsilon}{2^k}, \quad i = 1, \dots, k\,;\; r = a, b$$

where $\sharp_r(B_i)$ is the number of $X_r^{(j)}$, $j = 1, \dots n$, contained in the set B_i. Because of (5.3) there exists $P \in A_{a,b}$ in (5.2) such that $P << \overline{P}$. Hence (5.8) is a consequence of the law of large numbers which claims that $P^n(L_n \in A(\varepsilon, k))$ tends to 1 as n increases to infinity.

The conditioning of $\overline{\mathbf{P}}$ on the set $A(\varepsilon, k)$ in (5.7) by means of the empirical distribution L_n in (5.5) can now be defined as

$$\overline{\mathbf{P}}^{(n,k)}(\,.\,) = \overline{\mathbf{P}}(\,.\mid L_n \in A(\varepsilon, k)). \tag{5.9}$$

The dependence on $\varepsilon > 0$ is not indicated explicitly. It is purely technical and will be employed in Section 5.2, particularly in Lemma 5.2.2 and Lemma 5.2.9. We notice that the marginals of $\overline{\mathbf{P}}^{(n,k)}$ on Ω belong to $A(\varepsilon, k)$. In fact, the convex combination of elements in $A(\varepsilon, k)$

$$\int_{\Omega^n} L_n(\mathbf{X})(\,.\,)\, d\overline{\mathbf{P}}(\mathbf{X} \mid L_n \in A(\varepsilon, k)) \tag{5.10}$$

$$= \int_{\Omega^n} \frac{1}{n} \sum_{i=1}^n \delta_{X^{(i)}}(\,.\,)\, d\overline{\mathbf{P}}(\mathbf{X} \mid L_n \in A(\varepsilon, k))$$

$$= \frac{1}{n} \sum_{i=1}^n \overline{\mathbf{P}}(X^{(i)} \in \,.\mid L_n \in A(\varepsilon, k)) = \overline{\mathbf{P}}(X^{(j)} \in \,.\mid L_n \in A(\varepsilon, k))$$

is just the j-th marginal distribution of $\overline{\mathbf{P}}^{(n,k)}$ on Ω, $j = 1, \dots, n$. Lemma 5.2.3 will show that $A(\varepsilon, k)$ is completely convex, hence it contains the expression (5.10).

Let us state the main results of this chapter.

Theorem 5.1.1 *Let the subset $A_{a,b}$ of $M_1(\Omega)$ in (5.2) satisfy the assumption (5.3) and let $\varepsilon > 0$. Then:*

(i) $A(\varepsilon, k)$ defined in (5.7) approximates $A_{a,b}$ in the sense of

$$A_{a,b} = \bigcap_{k \in \mathbb{N}} A(\varepsilon, k). \tag{5.11}$$

Moreover, the set $A_{a,b}$ possesses the approximate Sanov property

$$\lim_{k \nearrow \infty} \lim_{n \nearrow \infty} \frac{1}{n} \log \overline{\mathbf{P}}(L_n \in A(\varepsilon, k)) = - \inf_{P \in A_{a,b}} H(P \mid \overline{P}) \tag{5.12}$$

where the infimum is attained by Csiszar's projection Q of $\overline{P}$ on the set $A_{a,b}$, i.e.,

$$\inf_{P \in A_{a,b}} H(P \mid \overline{P}) = H(Q \mid \overline{P}) \tag{5.13}$$

where $Q \in A_{a,b}$.

(ii) The system $(X^{(1)}, \ldots, X^{(n)})$ considered under the conditional proba-bility $\overline{\mathbf{P}}^{(n,k)}$ in (5.9) is weakly asymptotically quasi-independent *with limiting distribution Q, i.e.,*

$$\lim_{k \nearrow \infty} \lim_{n \nearrow \infty} \frac{1}{n} H(Q^n(\,.\mid L_n \in A(\varepsilon, k)) \mid \overline{\mathbf{P}}^{(n,k)}) = 0 \qquad (5.14)$$

where Q^n, which is the n-product of Csiszar's projection Q in (5.13), satisfies

$$\lim_{n \nearrow \infty} Q^n(L_n \in A(\varepsilon, k)) = 1 \qquad (5.15)$$

for all $k \in I\!N$.

A proof of the theorem will be given in the following section.

Remark 5.1.1 Csiszar (1984) calls $(\mathbf{P}^{(n)})_{n \in I\!N} \subset M_1(\Omega)$ asymptotically quasi-independent with limiting distribution $P \in M_1(\Omega)$, if it satisfies

$$\lim_{n \nearrow \infty} H(\mathbf{P}^{(n)} \mid P^n) = 0.$$

However, the product form of the limiting distribution P is actually only es-sential in the limit as n tends to infinity. This suggests the weak version of asymptotical quasi-independence introduced in (5.14) on the basis of (5.15). Additionally, our version deals with a reversed order of the arguments as it will be appropriate in our application.

If we would assume that there exists $P \in A_{a,b}$ with $H(P \mid \overline{P}) < \infty$ which is measure-theoretically equivalent to $\overline{P}$, then Csiszar's inequality (5.19) would imply that the Csiszar projection Q of $\overline{P}$ on $A_{a,b}$ in (5.13) is measure-theoretically equivalent to $\overline{P}$. As a consequence, $\overline{\mathbf{P}}^{(n,k)}$ would be absolutely continuous with respect to Q^n. However, it is just the intention of this chapter to consider situations in which Q^n vanishes on sets where $\overline{\mathbf{P}}^{(n,k)}$ does not vanish in general. In fact, Proposition 6.2.1 claims that certain Csiszar projections are Schrödinger processes. The zero set of their distribution is, as revealed in Chap-ter 2, a source of singular phenomena. We want to investigate them by means the large deviation principle (5.12) expressed in terms of the system $\overline{\mathbf{P}}^{(n,k)}$

Theorem 5.1.1 claims that Csiszar's projection Q is in $A_{a,b}$. Hence the law of large numbers yields $Q^n(L_n \in A(\varepsilon, k)) \to 1$ as n tends to infinity for $k \in I\!N$

because of (5.11). However, we have to be aware of $Q^n(L_n \in A(\varepsilon, k)) < 1$ in case of $n \in I\!\!N$. In order to get absolute continuity with respect to $\overline{\mathbf{P}}^{(n,k)}$, we condition in (5.14) the product Q^n on the set $\{L_n \in A(\varepsilon, k)\}$. Nevertheless, the product form reappears in the limit as n tends to infinity for $k \in I\!\!N$.

Let us briefly have a general discussion of our version of asymptotical quasi-independence of probability measures $\mathbf{P}^{(n)} \in M_1(\Omega^n)$, $n \in I\!\!N$, with limiting distribution $P \in M_1(\Omega)$. Provided

$$\lim_{n \nearrow \infty} \frac{1}{n} H(P^n \mid \mathbf{P}^{(n)}) = 0 \tag{5.16}$$

then any event holds asymptotically P^n-a.s. as n tends to infinity if its $\mathbf{P}^{(n)}$-probability tends to 1 with exponential rate. In fact, for any $\mathbf{B}_n \in \sigma(\Omega^n)$, the partition $\{\mathbf{B}_n, \mathbf{B}_n^c\}$ of Ω^n employed later in (5.40) yields

$$P^n(\mathbf{B}_n) \log(\frac{P^n(\mathbf{B}_n)}{\mathbf{P}^{(n)}(\mathbf{B}_n)}) + (1 - P^n(\mathbf{B}_n)) \log(\frac{1 - P^n(\mathbf{B}_n)}{1 - \mathbf{P}^{(n)}(\mathbf{B}_n)}) \le H(P^n \mid \mathbf{P}^{(n)})$$

where the left-hand side is non-negative. Divided by n, the right-hand side vanishes as n tends to infinity because of (5.16). Hence, if

$$\mathbf{P}^{(n)}(\mathbf{B}_n) = 1 - k_1 e^{-k_2 n}, \quad k_1, \ k_2 \ \text{positive constants}$$

then $\lim_{n \nearrow \infty} P^n(\mathbf{B}_n^c) = 0$.

5.2   Csiszar's projection and τ_0-topology

We prepare a series of lemmas for the proof of Theorem 5.1.1 which will follow at the end of this section.

Lemma 5.2.1 *Let* $A(\varepsilon, k)$ *be the subset of* $M_1(\Omega)$ *defined in (5.7). Then*

$$(i) \qquad A(\varepsilon, k) \ \text{decreases as} \ k \nearrow \infty$$

$$(ii) \qquad A(\varepsilon, k) \ \text{decreases as} \ \varepsilon \searrow 0$$

$$(iii) \qquad A_{a,b} = \bigcap_{k \in I\!\!N} A(\varepsilon, k), \quad \forall \varepsilon > 0$$

where $A_{a,b}$ *in (5.2) is assumed to satisfy (5.3).*

Proof. (i) Following (5.6), $\mathcal{P}_{k+1}(I\!\!R^d)$ is a refinement of $\mathcal{P}_k(I\!\!R^d)$. Hence, if $P \in A(\varepsilon, k+1)$ then $P \circ X_r^{-1} << \overline{P} \circ X_r^{-1}$ on $\sigma(\mathcal{P}_k(I\!\!R^d))$ for $r = a, b$. Moreover, every $B \in \mathcal{P}_k(I\!\!R^d)$ is of the form $B = B_1 \cup B_2$ with $B_1, B_2 \in \mathcal{P}_{k+1}(I\!\!R^d)$ and we can estimate

$$| P(X_r \in B_1 \cup B_2) - q_r(B_1 \cup B_2) |$$
$$\leq \quad | P(X_r \in B_1) - q_r(B_1) | + | P(X_r \in B_2) - q_r(B_2) |$$
$$\leq \quad \frac{\varepsilon}{2^{k+1}} + \frac{\varepsilon}{2^{k+1}} = \frac{\varepsilon}{2^k}$$

for $r = a, b$. Hence $P \in A(\varepsilon, k)$ which implies $A(\varepsilon, k+1) \subset A(\varepsilon, k)$.

(ii) is an immediate consequence of the definition of $A(\varepsilon, k)$ in (5.7).

(iii) The inclusion '$\subset$' follows from

$$A_{a,b} \subset A(\varepsilon, k), \quad \forall \varepsilon > 0, \ \forall k \in I\!\!N.$$

In fact, (5.3) provides $\tilde{P} \in A_{a,b}$ in (5.2) with $H(\tilde{P} \mid \overline{P}) < \infty$. Hence we have $\tilde{P} << \overline{P}$ on $\sigma(\Omega)$ and $q_r = \tilde{P} \circ X_r^{-1} << \overline{P} \circ X_r^{-1}$ on $\sigma(I\!\!R^d)$ for $r = a, b$. Thus $P \circ X_r^{-1} << \overline{P} \circ X_r^{-1}$ on $\sigma(\mathcal{P}_k(I\!\!R^d))$ follows for $r = a, b$ and for all $P \in A_{a,b}$.

To prove the converse inclusion '$\supset$', let us take $P \in \bigcap_{k \in I\!\!N} A(\varepsilon, k)$ for any $\varepsilon > 0$ and suppose that $P \notin A_{a,b}$. Hence there are $\varepsilon_0 > 0$ and a subset $B \in \sigma(I\!\!R^d)$ which satisfy

$$| P(X_r \in B) - q_r(B) | > \varepsilon_0 \quad \text{for} \ \ r = a \ \text{or} \ b.$$

Because of $\sigma(\mathcal{P}_k(I\!\!R^d)) \nearrow \sigma(I\!\!R^d)$ as $k \nearrow \infty$ in (5.6), we can find $k_0 \in I\!\!N$ and $B_i \in \mathcal{P}_k(I\!\!R^d)$ such that

$$| P(X_r \in B_i) - q_r(B_i) | > \frac{\varepsilon_0}{2^{k_0}} \quad \text{for} \ \ r = a \ \text{or} \ b$$

which is a contradiction. $\Diamond$

Lemma 5.2.2 *Let $\mathring{A}(\varepsilon, k)$ denote the interior of the set $A(\varepsilon, k)$ in (5.7) with respect to the τ_0-topology in (5.1). Then*

$$\mathring{A}(\varepsilon_0, k) \supset \bigcup_{0 < \varepsilon < \varepsilon_0} A(\varepsilon, k) \tag{5.17}$$

for every $\varepsilon_0 > 0$.

Proof. Let $\varepsilon_1 \in (0, \varepsilon_0)$ such that $P_1 \in A(\varepsilon_1, k)$. For $B_i \in \mathcal{P}_k(\mathbb{R}^d)$, $i = 1, \ldots, k$, we set

$$\Omega_i^{(r)} = \{X \in \Omega : X_r \in B_i\} \quad \text{for} \quad r = a, b$$

and obtain a finite measurable partition $\mathcal{P}$ of Ω by

$$\mathcal{P} = \{\Omega_i^{(a)} \cap \Omega_j^{(b)} : i, j = 1, \ldots, k\}.$$

Let $P \in U_0(P_1, c_2, \mathcal{P})$. Hence $P \ll P_1$ on $\sigma(\mathcal{P})$ and $P_1 \circ X_r^{-1} \ll \overline{P} \circ X_r^{-1}$ on $\sigma(\mathcal{P}_k(\mathbb{R}^d))$ for $r = a, b$. We conclude that $P \circ X_r^{-1} \ll \overline{P} \circ X_r^{-1}$ on $\sigma(\mathcal{P}_k(\mathbb{R}^d))$ for $r = a, b$ and that

$$\begin{aligned}
&\mid P(X_r \in B_i) - q_r(B_i) \mid \\
\leq \; &\mid P(\Omega_i^{(r)}) - P_1(\Omega_i^{(r)}) \mid + \mid P_1(X_r \in B_i) - q_r(B_i) \mid \\
\leq \; &k\varepsilon_2 + \frac{\varepsilon_1}{2^k}
\end{aligned}$$

for $r = a, b$ and $B_i \in \mathcal{P}_k(\mathbb{R}^d)$, $i = 1, \ldots k$. Consequently, if $\varepsilon_2 > 0$ is such that $k\varepsilon_2 + \varepsilon_1/2^k \leq \varepsilon_0/2^k$, then $P \in A(\varepsilon_0, k)$. Thus $U_0(P_1, \varepsilon_2, \mathcal{P}) \subset A(\varepsilon_0, k)$ which implies (5.17). $\Diamond$

Lemma 5.2.3 *(i) Let $(\Lambda, \mathcal{A}, \mu)$ be any probability space and let $\eta(\lambda, B)$ be a probability kernel on $\Lambda \times \sigma(\Omega)$ such that $\eta(\lambda, \, .\,) \in A(\varepsilon, k)$, $\forall \lambda \in \Lambda$. Then*

$$\mu[\eta] \equiv \int \eta(\lambda, \, .\,) \, d\mu(\lambda) \in A(\varepsilon, k)$$

i.e., the set $A(\varepsilon, k)$ in (5.7) is completely convex *in the sense of Csiszar (1984) and hence convex.*

(ii) Let $(P_n)_{n \in \mathbb{N}} \subset A(\varepsilon, k)$ be a sequence which converges to $P \in M_1(\Omega)$ in variation with respect to $\overline{P}$, i.e.,

$$\mid P_n - P \mid_{var:\overline{P}} \; = \int \mid \frac{dP_n}{d\overline{P}} - \frac{dP}{d\overline{P}} \mid d\overline{P} \to 0 \quad as \; n \nearrow \infty.$$

Then $P \in A(\varepsilon, k)$, i.e., $A(\varepsilon, k)$ is variation closed.

Proof. (i) As a consequence of (5.7),

$$\mid \mu[\eta](X_r \in B_i) - q_r(B_i) \mid \leq \int \mid \eta(\lambda, \{X_r \in B_i\}) - q_r(B_i) \mid d\mu(\lambda) \leq \frac{\varepsilon}{2^k}$$

for $r = a, b$ and $B_i \in \mathcal{P}_k(\mathbb{R}^d)$, $i = 1, \ldots, k$. Moreover, $\mu[\eta] \circ X_r^{-1} << \overline{P} \circ X_r^{-1}$ on $\sigma(\mathcal{P}_k(\mathbb{R}^d))$ for $r = a, b$. Hence $\mu[\eta] \in A(\varepsilon, k)$ follows.

(ii) For every $B_i \in \mathcal{P}_k(\mathbb{R}^d)$, $i = 1, \ldots, k$, and $r = a, b$,

$$
\begin{aligned}
| P(X_r \in B_i) - q_r(B_i) | \ &\leq \ | P(X_r \in B_i) - P_n(X_r \in B_i) | \\
&+ \ | P_n(X_r \in B_i) - q_r(B_i) |
\end{aligned}
$$

where the first member of the right-hand side satisfies

$$
| P(X_r \in B_i) - P_n(X_r \in B_i) | \leq | P_n - P |_{var:\overline{P}} \to 0 \quad \text{as } n \nearrow \infty
$$

and the second member of the right-hand side can be estimated as

$$
| P_n(X_r \in B_i) - q_r(B_i) | \leq \frac{\varepsilon}{2^k}
$$

because of $P_n \in A(\varepsilon, k)$ for $n \in \mathbb{N}$. Hence,

$$
| P(X_r \in B_i) - q_r(B_i) | \leq \frac{\varepsilon}{2^k} \quad \text{for } i = 1, \ldots, k; \ r = a, b
$$

which implies $P \in A(\varepsilon, k)$. $\diamondsuit$

For later reference we formulate Csiszar (1975)'s theorems 2.1 and 2.2 as Lemma 5.2.4. Lemma 5.2.5 and Lemma 5.2.6 relate convergence in entropy to convergence in variational distance and weak convergence, respectively. A detailed investigation of the relative entropy based on Csiszar (1975, 1984)'s contributions can be found in Nagasawa (1993)'s chap. X and XI.

Lemma 5.2.4 (Csiszar 1975) *Let* $(\Omega, \sigma(\Omega))$ *be any measurable space and let* $\overline{P} \in M_1(\Omega)$ *be fixed. If a subset A of $M_1(\Omega)$ is convex as well as variation closed and if it contains at least one element P with $H(P \mid \overline{P}) < \infty$, then there exists a unique I-projection $Q \in A$ of $\overline{P}$ on A defined by*

$$
\inf_{P \in A} H(P \mid \overline{P}) = H(Q \mid \overline{P}) \tag{5.18}
$$

where

$$
H(P \mid \overline{P}) \geq H(P \mid Q) + H(Q \mid \overline{P}), \quad \forall P \in A. \tag{5.19}
$$

From now on we will call the I-projection Csiszar's projection *and the relation (5.19)* Csiszar's inequality.

Lemma 5.2.5 (Csiszar 1967, Kemperman 1967, Kullback 1967)

The variational distance is dominated by the relative entropy according to

$$| P - R |_{var:R} \le \sqrt{2H(P \mid R)} \qquad (5.20)$$

for any P, $R \in M_1(\Omega)$.

Proof. We may evidently assume that $P << R$. Let

$$h(x) = 4 + 2x$$

and

$$g(x) = x \log x - x + 1$$

for $x > 0$. Both are non-negative and they satisfy

$$3(x - 1)^2 \le h(x)\, g(x). \qquad (5.21)$$

In fact, $h(x)\, g(x) - 3(x - 1)^2$ is convex and takes its minimal value 0 at $x = 1$. Thus (5.21) evaluated along dP/dR yields

$$3[\int | \frac{dP}{dR} - 1 | \, dR]^2 = [\int \sqrt{3 \, | \frac{dP}{dR} - 1 |^2} \, dR]^2$$

$$\le \int h(\frac{dP}{dR}) \, dR \int g(\frac{dP}{dR}) \, dR \le 6 \int g(\frac{dP}{dR}) \, dR$$

$$= 6 \int \frac{dP}{dR} \, \log(\frac{dP}{dR}) \, dR.$$

$\Diamond$

Lemma 5.2.6 (Csiszar 1975) *If $\lim_{n \nearrow \infty} H(P_n \mid P) = 0$ for $P, P_n \in M_1(\Omega)$, $n \in I\!N$, then*

$$\lim_{n \nearrow \infty} \int f \, dP_n = \int f \, dP$$

provided that f is any measurable function for which $\exp\{t\, f\}$ is P-integrable if $| t |$ is sufficiently small.

As a consequence, convergence in entropy implies weak convergence.

Proof. The P-density of P_n denoted by p_n exists because of $H(P_n \mid P) < \infty$. Hence (5.20) yields

$$| P_n - P |_{var:P} = \int | p_n - 1 | \, dP \to 0 \quad \text{as } n \nearrow \infty$$

from which

$$\int_{\{|f|\leq M\}} f \, dP_n \to \int_{\{|f|\leq M\}} f \, dP \quad \text{as } n \nearrow \infty$$

follows for any constant M. Consequently, for arbitrary $\delta > 0$, a constant M has to be found such that

$$\limsup_{n\nearrow\infty} \int_{\{|f|>M\}} |f| \, dP_n = \int_{\{|f|>M\}} |f| \, p_n \, dP < \delta. \tag{5.22}$$

We notice that $\lim_{n\nearrow\infty} H(P_n \mid P) = 0$ implies

$$\liminf_{n\nearrow\infty} \int_B p_n \log p_n \, dP = \limsup_{n\nearrow\infty} \int_B p_n \log p_n \, dP = 0 \tag{5.23}$$

for any $B \in \sigma(\Omega)$. In fact, because of $p_n \log p_n \geq e^{-1}$ for $n \in \mathbb{N}$, Fatou's lemma yields

$$\liminf_{n\nearrow\infty} \int 1_B \, p_n \, \log p_n \, dP \geq \int \liminf_{n\nearrow\infty}(1_B \, p_n \, \log p_n) \, dP = 0$$

and hence in particular,

$$\limsup_{n\nearrow\infty} \int 1_B \, p_n \, \log p_n \, dP$$

$$= \lim_{n\nearrow\infty} H(P_n \mid P) - \liminf_{n\nearrow\infty} \int 1_{\Omega\setminus B} \, p_n \, \log p_n \, dP = 0.$$

By assumption

$$\int \exp\{t \, f\} \, dP = P(\{f = 0\})$$

$$+ \int_{\{f^+>0\}} \exp\{t \, f^+\} \, dP + \int_{\{f^->0\}} \exp\{-t \, f^-\} \, dP$$

is finite for $|t|$ sufficiently small. Thus

$$\int \exp\{t \, |f|\} \, dP = P(\{f = 0\})$$

$$+ \int_{\{f^+>0\}} \exp\{t \, f^+\} \, dP + \int_{\{f^->0\}} \exp\{t \, f^-\} \, dP$$

is finite in case of sufficiently small $|t|$. As a consequence, there exist $t > 0$ and $M < \infty$ such that

$$\int_{\{|f|>M\}} \exp\{t \, |f|\} \, dP < \delta \, t.$$

In order to receive (5.22) we apply the inequality

$$a\,b < a\,\log a + e^b \quad \text{for} \quad a, b \geq 0$$

in Beckenbach-Bellman (1961)'s section 15 to $a_n = p_n$, $n \in I\!N$, and $b = t \mid f \mid$.
In fact, (5.22) equals

$$\frac{1}{t} \limsup_{n \nearrow \infty} \int_{\{|f|>M\}} a_n\, b\, dP < \frac{1}{t} \limsup_{n \nearrow \infty} \int_{\{|f|>M\}} (a_n \log a_n + e^b)\, dP$$

$$= \frac{1}{t} \limsup_{n \nearrow \infty} \left(\int_{\{|f|>M\}} a_n \log a_n\, dP + \int_{\{|f|>M\}} e^b\, dP \right) < \delta$$

by an application of (5.23) to $B = \{\mid f \mid > M\}$. $\Diamond$

In view of simplicity, let us employ the notation

$$H(A \mid \overline{P}) = \inf_{P \in A} H(P \mid \overline{P}) \tag{5.24}$$

for subsets $A \subset M_1(\Omega)$.

Lemma 5.2.7 ('Sanov relation') *Let $\varepsilon > 0$ and $k \in I\!N$. If (5.3) is provided, then*

$$-H(\mathring{A}(\varepsilon, k) \mid \overline{P}) \leq \liminf_{n \nearrow \infty} \frac{1}{n} \log \overline{\mathbf{P}}(L_n \in A(\varepsilon, k))$$

$$\tag{5.25}$$

$$\leq \limsup_{n \nearrow \infty} \frac{1}{n} \log \overline{\mathbf{P}}(L_n \in A(\varepsilon, k)) \leq -H(A(\varepsilon, k) \mid \overline{P})$$

*where $H(A(\varepsilon, k) \mid \overline{P})$ is attained by Csiszar's projection $Q_{\varepsilon,k}$ in (5.18) of $\overline{P}$ on
the set $A(\varepsilon, k)$ in (5.7), i.e.,*

$$H(A(\varepsilon, k) \mid \overline{P}) = H(Q_{\varepsilon,k} \mid \overline{P}). \tag{5.26}$$

Remark 5.2.1 The upper bound in the Sanov relation (5.25) can be refined by

$$\frac{1}{n} \log \overline{\mathbf{P}}(L_n \in A(\varepsilon, k)) \leq -\frac{1}{n} H(\overline{\mathbf{P}}^{(n,k)} \mid Q_{\varepsilon,k}^n) - H(Q_{\varepsilon,k} \mid \overline{P}) \tag{5.27}$$

for n, $k \in I\!N$, which corresponds to Csiszar (1984)'s (2.17). Let $\overline{\mathbf{P}}_1^{(n,k)}$ denote
the identical one-dimensional marginal distributions of $\overline{\mathbf{P}}^{(n,k)}$ in (5.10). They
satisfy

$$H(\overline{\mathbf{P}}_1^{(n,k)} \mid Q_{\varepsilon,k}) \leq \frac{1}{n} H(\overline{\mathbf{P}}^{(n,k)} \mid Q_{\varepsilon,k}^n) \tag{5.28}$$

where $Q_{\varepsilon,k}$ is Csiszar's projection of $\overline{P}$ on $A(\varepsilon, k)$ in (5.26). If $A(\varepsilon, k)$ possesses the Sanov property , i.e., (5.25) holds with equality, then (5.27) relates the speed of convergence in the Sanov property to the speed of convergence in entropy of the conditional probability $\overline{\mathbf{P}}_1^{(n,k)}$ to $Q_{\varepsilon,k}$.

Proof. Let $\mathbf{P}^{(n)} \in M_1(\Omega^n)$ with the marginals $P_i \in M_1(\Omega)$ and let $Q_i \in M_1(\Omega)$ such that $P_i << Q_i$ for $i = 1, \ldots, n$. Then

$$H(\mathbf{P}^{(n)} \mid \prod_{i=1}^{n} Q_i) = H(\mathbf{P}^{(n)} \mid \prod_{i=1}^{n} P_i) + \sum_{i=1}^{n} H(P_i \mid Q_i). \tag{5.29}$$

In fact, following Csiszar (1984)'s section 2, we have

$$\frac{d\mathbf{P}^{(n)}}{d\prod_{i=1}^{n} Q_i}(\mathbf{X}) = \frac{d\mathbf{P}^{(n)}}{d\prod_{i=1}^{n} P_i}(\mathbf{X}) \prod_{i=1}^{n} \frac{dP_i}{dQ_i}(X^{(i)})$$

which yields (5.29) by taking logarithm and integrating with respect to $\mathbf{P}^{(n)}$. Because of the non-negativity of the relative entropy, (5.29) implies

$$H(P \mid Q) \leq \frac{1}{n} H(\mathbf{P}^{(n)} \mid Q^n) \tag{5.30}$$

if all marginals of $\mathbf{P}^{(n)}$ equal P and $Q = Q_i$ for $i = 1, \ldots, n$. We receive (5.28) in case of $\mathbf{P}^{(n)} = \overline{\mathbf{P}}^{(n,k)}$ and $Q = Q_{\varepsilon,k}$.

In view of the upper bound in (5.25), we consider $H(\overline{\mathbf{P}}^{(n,k)} \mid \overline{P}^n)$. Because of (5.9)

$$\frac{d\overline{\mathbf{P}}^{(n,k)}}{d\overline{P}^n} = \frac{1}{\overline{\mathbf{P}}(L_n \in A(\varepsilon,k))} \delta_{\{L_n \in A(\varepsilon,k)\}}$$

and hence

$$H(\overline{\mathbf{P}}^{(n,k)} \mid \overline{P}^n) = -\log \overline{\mathbf{P}}(L_n \in A(\varepsilon,k)). \tag{5.31}$$

In case of $P^{(n)} = \overline{\mathbf{P}}^{(n,k)}$, $P_i = \overline{\mathbf{P}}_1^{(n,k)}$ and $Q_i = \overline{P}$ for $i = 1, \ldots, n$, (5.29) means

$$H(\overline{\mathbf{P}}^{(n,k)} \mid \overline{P}^n) = H(\overline{\mathbf{P}}^{(n,k)} \mid (\overline{\mathbf{P}}_1^{(n,k)})^n) + n\,H(\overline{\mathbf{P}}_1^{(n,k)} \mid \overline{P}).$$

Hence (5.31) yields

$$-\log \overline{\mathbf{P}}(L_n \in A(\varepsilon,k)) = H(\overline{\mathbf{P}}^{(n,k)} \mid (\overline{\mathbf{P}}_1^{(n,k)})^n) + n\,H(\overline{\mathbf{P}}_1^{(n,k)} \mid \overline{P}). \tag{5.32}$$

Following Lemma 5.2.1, $A_{a,b}$ is contained in $A(\varepsilon, k)$. Hence there is at least one $P \in A(\varepsilon, k)$ such that $H(P \mid \overline{P}) < \infty$ because of assumption (5.3). Lemma 5.2.3

claims that $A(\varepsilon, k)$ is convex and variation closed. Consequently, Lemma 5.2.4 can be employed to receive Csiszar's projection $Q_{\varepsilon,k}$ of $\overline{P}$ on $A(\varepsilon, k)$. With $Q_i = Q_{\varepsilon,k}$ for $i = 1, \ldots, n$, (5.29) becomes

$$H(\overline{\mathbf{P}}^{(n,k)} \mid Q_{\varepsilon,k}^n) = H(\overline{\mathbf{P}}^{(n,k)} \mid (\overline{\mathbf{P}}_1^{(n,k)})^n) + n\, H(\overline{\mathbf{P}}_1^{(n,k)} \mid Q_{\varepsilon,k})$$

and hence (5.32) can be written as

$$-\log \overline{\mathbf{P}}(L_n \in A(\varepsilon, k)) = H(\overline{\mathbf{P}}^{(n,k)} \mid Q_{\varepsilon,k}^n) + n\,[H(\overline{\mathbf{P}}_1^{(n,k)} \mid \overline{P}) - H(\overline{\mathbf{P}}_1^{(n,k)} \mid Q_{\varepsilon,k})].$$

Since $\overline{\mathbf{P}}_1^{(n,k)} \in A(\varepsilon, k)$ as a consequence of (5.10), Csiszar's inequality (5.19) yields

$$H(\overline{\mathbf{P}}_1^{(n,k)} \mid \overline{P}) \geq H(\overline{\mathbf{P}}_1^{(n,k)} \mid Q_{\varepsilon,k}) + H(Q_{\varepsilon,k} \mid \overline{P})$$

and (5.27) follows. It provides the upper bound in (5.25) because of the non-negativity of the relative entropy.

In view of the lower bound in (5.25), i.e.,

$$-H(\mathring{A}(\varepsilon, k) \mid \overline{P}) \leq \liminf_{n \nearrow \infty} \frac{1}{n} \log \overline{\mathbf{P}}(L_n \in \mathring{A}(\varepsilon, k)) \tag{5.33}$$

assumption (5.3) ensures $H(\mathring{A}(\varepsilon, k) \mid \overline{P}) < \infty$ because of Lemma 5.2.1 and Lemma 5.2.2. In terms of (5.24) there exists $P \in \mathring{A}(\varepsilon, k)$ such that

$$H(P \mid \overline{P}) < H(\mathring{A}(\varepsilon, k) \mid \overline{P}) + \delta \tag{5.34}$$

for arbitrary $\delta > 0$. Lemma 5.2.8 will show that the set of finite atomar probability measures $M_1^{f.a.}(\Omega)$ is open in the τ_0-topology. Consequently, there exists a finite measurable partition $\mathcal{P} = \{\Omega_1, \ldots, \Omega_k\}$ of Ω and $\varepsilon > 0$ such that

$$U_0(P, \varepsilon, \mathcal{P}) \cap M_1^{f.a.}(\Omega) \subset \mathring{A}(\varepsilon, k).$$

Because of $H(P \mid \overline{P}) < \infty$, $P << \overline{P}$ on $\sigma(\Omega)$. Hence there are $0 < \varepsilon_1 < \varepsilon$ and non-negative rational numbers $r_1, \ldots, r_k$ with

$$\mid r_i - P(\Omega_i) \mid < \varepsilon_1 \quad \text{and} \quad r_i = 0 \quad \text{if} \quad P(\Omega_i) = 0, \quad i = 1, \ldots, k \tag{5.35}$$

such that

$$\mid r_i \log \frac{r_i}{\overline{P}(\Omega_i)} - P(\Omega_i) \log \frac{P(\Omega_i)}{\overline{P}(\Omega_i)} \mid < \frac{\delta}{k}, \quad i = 1, \ldots, k. \tag{5.36}$$

Thus

$$\overline{\mathbf{P}}(L_n \in \mathring{A}(\varepsilon, k)) \geq \overline{\mathbf{P}}(L_n \in U_0(P, \mathcal{P}, \varepsilon) \cap M_1^{f.a.}(\Omega)) \qquad (5.37)$$

$$\geq \overline{\mathbf{P}}(L_n(\mathbf{X})(\Omega_i) = r_i, \; i = 1, \ldots, k)$$

$$= \overline{\mathbf{P}}(\frac{1}{n} \sum_{j=1}^{n} \delta_{X^{(j)}}(\Omega_i) = r_i, \; i = 1, \ldots, k)$$

follows. For sufficiently large n, we can find $l_i \in I\!\!N$, $i = 1, \ldots, k$, which satisfy $r_i = l_i/n$, $i = 1, \ldots, n$, and $\sum_{i=1}^{k} l_i = n$. Since $\mathcal{P} = \{\Omega_i, \; i = 1, \ldots, k\}$ is a finite measurable partition of Ω, $\mathbf{X} = (X^{(1)}, \ldots, X^{(n)})$ is in $\{\frac{1}{n} \sum_{j=1}^{n} \delta_{X^{(j)}}(\Omega_i) = r_i\}$ if and only if l_i of the $X^{(j)}$, $j = 1, \ldots, n$, are in Ω_i. Let us consider a certain choice of l_i-subsets in the n-set of components of $\mathbf{X}$. The a priori probability of the corresponding set of $\mathbf{X}$'s is

$$\prod_{i=1}^{k} \overline{P}(\Omega_i)^{l_i}.$$

The number of choices to arrange these l_i-subsets in the n-set of $X^{(j)}$'s is

$$\frac{n!}{l_1! \, (n - l_1)!} \, \frac{(n - l_1)!}{l_2! \, (n - l_1 - l_2)!} \cdot \ldots \cdot \frac{(n - l_1 - \ldots - l_{k-1})!}{l_k! \, 0!} = \frac{n!}{l_1! \cdot \ldots \cdot l_k!}.$$

Consequently, (5.37) provides the estimate

$$\overline{\mathbf{P}}(L_n \in \mathring{A}(\varepsilon, k)) \geq \frac{n!}{l_1! \cdot \ldots \cdot l_k!} \prod_{i=1}^{k} \overline{P}(\Omega_i)^{l_i}. \qquad (5.38)$$

The right-hand side of (5.38) is now treated by means of Stirling's formula

$$\log(n!) = (n + \frac{1}{2}) \log n - n + \frac{1}{2} \log(2\pi) + \log(1 + \kappa_n)$$

where $\kappa_n \to 0$ as $n \nearrow \infty$. We obtain

$$\log\{\frac{n!}{l_1! \cdot \ldots \cdot l_k!} \prod_{i=1}^{k} \overline{P}(\Omega_i)^{l_i}\} \qquad (5.39)$$

$$= -n \sum_{i=1}^{k} r_i \log(\frac{r_i}{\overline{P}(\Omega_i)}) - \frac{1}{2k} \sum_{i=1}^{k} \log r_i + \frac{1-k}{2} \log(2\pi n)$$

$$+ \log(1 + \kappa_n) - \sum_{i=1}^{k} \log(1 + \kappa_{l_i})$$

$$\geq -n \sum_{i=1}^{k} r_i \log(\frac{r_i}{\overline{P}(\Omega_i)}) + \frac{1-k}{2} \log(2\pi n) + \log(1 + \kappa_n) - \sum_{i=1}^{k} \log(1 + \kappa_{l_i}).$$

The convex function $f(x) = x \log x$, $x \geq 0$, is going to provide an approximation of the relative entropy from below. In fact, if $\sigma(\mathcal{P})$ denotes the σ-algebra generated by the partition $\mathcal{P}$, then

$$H(P \mid \overline{P}) = \overline{P}[f(p)] = \overline{P}[\overline{P}[f(p) \mid \sigma(\mathcal{P})]]$$

for $p = dP/d\overline{P}$, where $P \in M_1(\Omega)$ with $P << \overline{P}$, and hence Jensen's inequality yields

$$f(\overline{P}[p \mid \sigma(\mathcal{P})]) \leq \overline{P}[f(p) \mid \sigma(\mathcal{P})], \quad \overline{P}\text{-a.s.}$$

Because of

$$f(\overline{P}[p \mid \sigma(\mathcal{P})]) = \sum_{i=1}^{k} f\left(\frac{P(\Omega_i)}{\overline{P}(\Omega_i)}\right) 1_{\Omega_i}, \quad \overline{P}\text{-a.s.}$$

we receive

$$\sum_{i=1}^{k} P(\Omega_i) \log\left(\frac{P(\Omega_i)}{\overline{P}(\Omega_i)}\right) \leq H(P \mid \overline{P}) \tag{5.40}$$

where (5.36) provides

$$\left| \sum_{i=1}^{k} r_i \log\left(\frac{r_i}{\overline{P}(\Omega_i)}\right) - \sum_{i=1}^{k} P(\Omega_i) \log\left(\frac{P(\Omega_i)}{\overline{P}(\Omega_i)}\right) \right| < \delta.$$

As a consequence, (5.34) and (5.40) yield

$$\begin{aligned}
\sum_{i=1}^{k} r_i \log\left(\frac{r_i}{\overline{P}(\Omega_i)}\right) &\leq \sup_{\mathcal{P}} \sum_{i=1}^{k} P(\Omega_i) \log\left(\frac{P(\Omega_i)}{\overline{P}(\Omega_i)}\right) + \delta \\
&\leq H(P \mid \overline{P}) + \delta \leq H(\mathring{A}(\varepsilon, k) \mid \overline{P}) + 2\delta.
\end{aligned}$$

The lower bound in (5.25) follows now from (5.38) and (5.39) by means of

$$\liminf_{n \nearrow \infty} \frac{1}{n} \log \overline{\mathbf{P}}(L_n \in \mathring{A}(\varepsilon, k))$$

$$\geq \liminf_{n \nearrow \infty} \left\{ \frac{1-k}{2n} \log(2\pi n) - \sum_{i=1}^{k} r_i \log\left(\frac{r_i}{\overline{P}(\Omega_i)}\right) \right\}$$

$$\geq -H(\mathring{A}(\varepsilon, k) \mid \overline{P}) - 2\delta.$$

Remark 5.2.2 Let us consider Csiszar's (1984) τ_0-topology in some details. Groeneboom-Oosterhoff-Ruymgaart (1979) formulate in their lemma 3.1 the

Sanov relation (5.25) with respect to the τ-*topology* of $M_1(\Omega)$. This topology is defined in terms of the basic neighborhoods

$$U(P, \mathcal{P}, \varepsilon) = \{R \in M_1(\Omega) : \mid R(\Omega_i) - P(\Omega_i) \mid < \varepsilon, \ i = 1, \ldots, k\}$$

for $P \in M_1(\Omega)$ where $\mathcal{P}$ is any finite measurable partition of Ω and $\varepsilon > 0$. Hence the τ-topology is the coarsest topology of $M_1(\Omega)$ in which the mappings

$$R \in M_1(\Omega) \to R(\mathbf{B}) \in [0, \infty)$$

are continuous for sets $\mathbf{B} \in \sigma(\Omega)$. Moreover, $(P_n)_{n \in \mathbb{N}} \subset M_1(\Omega)$ converges to $P \in M_1(\Omega)$ in the τ-topology,

$$P_n \xrightarrow{\tau} P \iff \lim_{n \nearrow \infty} \int_\Omega f \, dP_n = \int_\Omega f \, dP$$

$\forall f : \Omega \to \mathbb{R}$, bounded and $\sigma(\Omega)$-measurable. Consequently, the so-called weak topology generated by the Lévy-Prohorov metric which requires test functions $f \in C_b(\Omega)$ is coarser than the τ-topology.

The τ_0-topology is generated by basic neighborhoods of $P \in M_1(\Omega)$ defined in (5.1). We notice that $P_n \xrightarrow{\tau_0} P$ is not equivalent to $\int f \, dP_n \to \int f \, dP$, $\forall f : \Omega \to \mathbb{R}$, bounded and $\sigma(\Omega)$-measurable. In fact, $P_n \xrightarrow{\tau_0} P$ claims that $\forall \varepsilon > 0$ and $\forall \mathcal{P} \subset \sigma(\Omega)$ there is $n_\varepsilon \in \mathbb{N}$ such that $\forall n \geq n_\varepsilon : P_n \in U_0(P, \mathcal{P}, \varepsilon)$, i.e., $\mid P_n(\Omega_i) - P(\Omega_i) \mid < \varepsilon, \ i = 1, \ldots, k$, and $P_n(\Omega_i) = 0$ if $P(\Omega_i) = 0$, $i = 1, \ldots, k$. Weak convergence however means only that $\mid P_n(\Omega_i) - P(\Omega_i) \mid < \varepsilon$. Consequently, if $P(\Omega_i) = 0$ then $\mid P_n(\Omega_i) \mid \leq \mid P_n(\Omega_i) - P(\Omega_i) \mid + \mid P(\Omega_i) \mid < \varepsilon$ which does not imply absolute continuity.

The absolute-continuity properties naturally required by finite relative entropy conditions show that the τ_0-topology might be more adequate in our situation than the τ-topology. However, we have to be aware that the τ_0-topology is coarser than the τ-topology, hence the Sanov relation (5.25) is a weaker statement in the τ_0-topology than in the τ-topology. In fact, the interior of a set with respect to the τ_0-topology is contained in that with respect to the τ-topology, hence the lower bound in (5.25) is smaller in the τ_0-topology than in the τ-topology. Moreover, the closure of a set with respect to the τ_0-topology contains the closure of this set with respect to the τ-topology, hence the upper bound in (5.25) is bigger in the τ_0-topology than in the τ-topology. However, the τ_0-topology enables us to prove the approximate Sanov property in (5.12) which itself does not refer to a particular topology. Actually, Theorem 5.1.1

claims that the Sanov relation (5.25) holds with equality in the limit as n and k tend to infinity.

Lemma 5.2.8 *The set of finite atomar probability measures defined as*

$$M_1^{f.a.}(\Omega) \;=\; \{P \in M_1(\Omega) : P = \sum_{i=1}^{k} \alpha_i\, \delta_{X^{(i)}} \;\; where$$

$$X^{(i)} \in \Omega,\; \alpha_i > 0,\; i = 1, \ldots, k, \;\; and \;\; \sum_{i=1}^{k} \alpha_i = 1,\; k \in \mathbb{N} \,\}$$

is τ_0-open but not τ-open.

Proof. (*i*) For each $P \in M_1^{f.a.}(\Omega)$ represented as $P = \sum_{i=1}^{k} \alpha_i\, \delta_{X^{(i)}}$ we have to find $\varepsilon > 0$ and $\mathcal{P} \subset \sigma(\Omega)$ such that $U_0(P, \mathcal{P}, \varepsilon) \subset M_1^{f.a.}(\Omega)$. The choice $\mathcal{P} = \{\Omega_1 = \{X^{(1)}, \ldots, X^{(k)}\}, \; \Omega_2 = \Omega\backslash\Omega_1\}$ yields $P(\Omega_1) = 1$ and $P(\Omega_2) = 0$. Because of $R << P$ on $\sigma(\mathcal{P})$ for every $R \in U_0(P, \mathcal{P}, \varepsilon)$, $R(\Omega_2) = 0$, i.e., $R(\Omega_2^c) = R(\Omega_1) = 1$. Hence there are $\beta_i \geq 0$, $i = 1, \ldots, k$, with $\sum_{i=1}^{k} \beta_i = 1$ such that $R = \sum_{i=1}^{k} \beta_i\, \delta_{X^{(i)}}$, i.e., $R \in M_1^{f.a.}(\Omega)$. Because of $R(\Omega_i) = P(\Omega_i)$, $i = 1, 2$, for each such R, $U_0(P, \mathcal{P}, \varepsilon) \subset M_1^{f.a.}(\Omega)$, $\forall \varepsilon > 0$.

(*ii*) To deal with any other than only finite atomar probability measures, Ω needs to have infinitely many elements. For $P = \sum_{i=1}^{k} \alpha_i\, \delta_{X^{(i)}} \in M_1^{f.a.}(\Omega)$ and any finite partition $\mathcal{P} \subset \sigma(\Omega)$, there always exists $R \in M_1(\Omega)\backslash M_1^{f.a.}(\Omega)$ such that $R(\Omega_i) = P(\Omega_i)$, $\forall\, \Omega_i \in \mathcal{P}$. This means that $R \in U(P, \mathcal{P}, \varepsilon)$, $\forall \varepsilon > 0$. $\Diamond$

Lemma 5.2.9 *If (5.3) is provided, then:*

(*i*) *There exists Csiszar's projection Q of $\overline{P}$ on $A_{a,b}$ in (5.2), i.e.,*

$$H(A_{a,b} \mid \overline{P}) = H(Q \mid \overline{P}). \tag{5.41}$$

(*ii*) *The sequence $(A(\varepsilon, k))_{k \in \mathbb{N}}$ in (5.7) which approximates $A_{a,b}$ according to (5.11) satisfies*

$$\lim_{k \nearrow \infty} H(\mathring{A}(\varepsilon, k) \mid \overline{P}) = \lim_{k \nearrow \infty} H(A(\varepsilon, k) \mid \overline{P}) = H(Q \mid \overline{P}), \quad \forall \varepsilon > 0. \tag{5.42}$$

(*iii*) *Csiszar's projection $Q_{\varepsilon,k}$ of $\overline{P}$ on $A(\varepsilon, k)$ in (5.26) converges to Csiszar's projection Q of $\overline{P}$ on $A_{a,b}$ in (5.41) in entropy and hence weakly as k tends to infinity.*

Proof. Following (iii) of Lemma 5.2.1 and Lemma 5.2.3, the set $A_{a,b}$ in (5.2) is convex and variation closed. Provided (5.3), there exists Csiszar's projection Q of $\overline{P}$ on $A_{a,b}$ in (5.41) as a consequence of Lemma 5.2.4. For $0 < \varepsilon_1 < \varepsilon_2$, Lemma 5.2.1 and Lemma 5.2.2 yield in terms of (5.24)

$$H(\mathring{A}(\varepsilon_1, k) \mid \overline{P}) \;\geq\; H(A(\varepsilon_1, k) \mid \overline{P}) \tag{5.43}$$
$$\geq\; H(\mathring{A}(\varepsilon_2, k) \mid \overline{P}) \;\geq\; H(A(\varepsilon_2, k) \mid \overline{P})$$

as well as

$$H(Q \mid \overline{P}) \;=\; H(A_{a,b} \mid \overline{P}) \tag{5.44}$$
$$\geq\; H(A(\varepsilon, k) \mid \overline{P}) \;=\; H(Q_{\varepsilon,k} \mid \overline{P}), \quad \forall \varepsilon > 0$$

where (5.26) and (5.41) have been employed. The limit (5.42) is a consequence of the *lower semi-continuity* of the relative entropy deduced in (5.40). In fact, Lemma 5.2.1 implies

$$\lim_{k \nearrow \infty} H(A(\varepsilon, k) \mid \overline{P}) = H(A_{a,b} \mid \overline{P}), \quad \forall \varepsilon > 0 \tag{5.45}$$

from which

$$\lim_{k \nearrow \infty} H(\mathring{A}(\varepsilon, k) \mid \overline{P}) = H(A_{a,b} \mid \overline{P}), \quad \forall \varepsilon > 0 \tag{5.46}$$

follows because of (5.43).

Part (*iii*) is obtained from Csiszar's inequality (5.19) which claims that

$$H(Q \mid \overline{P}) - H(Q_{\varepsilon,k} \mid \overline{P}) \geq H(Q \mid Q_{\varepsilon,k}) \tag{5.47}$$

in case of $Q \in A_{a,b} \subset A(\varepsilon, k)$ in (5.11). Following (5.44) and (5.45), (5.47) yields

$$\lim_{k \nearrow \infty} H(Q \mid Q_{\varepsilon,k}) = 0, \quad \forall \varepsilon > 0 \tag{5.48}$$

i.e., $Q_{\varepsilon,k}$ converges to Q in entropy and hence weakly by Lemma 5.2.6. $\Diamond$

Proof of Theorem 5.1.1. (*i*) (5.11) is given in (*iii*) of Lemma 5.2.1 and (5.13) is provided by (5.41) in Lemma 5.2.9. In order to verify the approximate Sanov property (5.12), we estimate

$$\left| \frac{1}{n} \log \overline{\mathbf{P}}(L_n \in A(\varepsilon, k)) + H(Q \mid \overline{P}) \right|$$
$$\leq\; \left| \frac{1}{n} \log \overline{\mathbf{P}}(L_n \in A(\varepsilon, k)) + H(A(\varepsilon, k) \mid \overline{P}) \right|$$
$$+\; \left| -H(A(\varepsilon, k) \mid \overline{P}) + H(Q \mid \overline{P}) \right|$$

which vanishes as n and k tend to infinity by Lemma 5.2.7 and Lemma 5.2.9.

(ii) The limit (5.15) is a consequence of the law of large numbers because of (5.5), (5.7), (5.13) and (iii) of Lemma 5.2.1. The *Radon-Nikodym derivative*

$$\frac{dQ^n(\,.\,\mid L_n \in A(\varepsilon, k))}{d\overline{\mathbf{P}}^{(n,k)}}$$

exists because of (5.9) and (5.13) which implies $Q << \overline{P}$. A straightforward calculation yields

$$\frac{1}{n} H(Q^n(\,.\,\mid L_n \in A(\varepsilon, k)) \mid \overline{\mathbf{P}}^{(n,k)})$$

$$= \frac{1}{Q^n(L_n \in A(\varepsilon, k))} \frac{1}{n} \int_{\{L_n \in A(\varepsilon, k)\}} \log(\frac{dQ^n}{d\overline{P}^n})\, dQ^n$$

$$- \frac{1}{n} \log Q^n(L_n \in A(\varepsilon, k)) + \frac{1}{n} \log \overline{P}^n(L_n \in A(\varepsilon, k))$$

where the right-hand side vanishes as n and k tend to infinity. In fact, the second and third member of the right-hand side are treated in (5.12) and (5.15), respectively. The first member of the right-hand side can be represented as

$$\frac{1}{n} \sum_{i=1}^{n} \int_{\{L_n \in A(\varepsilon, k)\}} \log(\frac{dQ}{d\overline{P}})(X^{(i)})\, dQ^n(\mathbf{X})$$

$$= \int_{\Omega} \log(\frac{dQ}{d\overline{P}})(X^{(1)})\, dQ(X^{(1)})$$

$$\int_{\{(X^{(2)},\dots,X^{(n)}):(X^{(1)},X^{(2)},\dots,X^{(n)})\in\{L_n\in A(\varepsilon,k)\}\}} dQ^{n-1}(X^{(2)},\dots,X^{(n)})$$

which tends to $H(Q \mid \overline{P})$ as n tends to infinity for $k \in \mathbb{N}$ because of (5.3), (5.15) and the symmetry of L_n in (5.5) in its components $X^{(i)}$, $i = 1, \dots n$. $\diamondsuit$

Remark 5.2.3 We formulate Theorem 5.1.1 for probability measures on the space of continuous paths since we intend to apply it to diffusion processes. However, the continuity of paths plays no role and may be dropped. Let $(S, \sigma(S))$ be a measurable space with a countably generated σ-field $\sigma(S)$ and let Ω be the space of measurable functions on $[a, b]$ taking values in S. Then we consider the space $M_1(\Omega)$ of probability measures on this general paths space Ω and remark that all statements in Section 5.2 remain valid.

The conditional process $(\mathbf{X}_t = (X_t^{(1)}, \dots, X_t^{(n)})$, $a \le t \le b$, $\overline{\mathbf{P}}^{(n,k)})$ on Ω^n in (5.9) investigated in (ii) of Theorem 5.1.1 is in general not Markovian. However, the following Proposition 5.2.1 provides under the present assumptions a

Markovian modification $\mathbf{Q}^{(n,k)}$ of $\overline{\mathbf{P}}^{(n,k)}$ satisfying

$$H(\mathbf{Q}^{(n,k)} \mid Q^n_{\varepsilon,k}) \leq H(\overline{\mathbf{P}}^{(n,k)} \mid Q^n_{\varepsilon,k}). \tag{5.49}$$

In fact, we recall that our reference process $\overline{P}$ on Ω is assumed to be Markovian. Then Proposition 5.2.1 can be applied to $\mathbf{P} = Q$ and $\mathbf{P_0} = \overline{P}$ as well as to $\mathbf{P} = Q_{\varepsilon,k}$ and $\mathbf{P_0} = \overline{P}$ because of Lemma 5.2.9. It yields Markovian Csiszar projections which are again denoted by Q and $Q_{\varepsilon,k}$, respectively. Proposition 6.2.1 will show later that a modification is actually not even necessary. The Markovian modification $\mathbf{Q}^{(n,k)}$ follows now by an application of Proposition 5.2.1 to $\mathbf{P} = \overline{\mathbf{P}}^{(n,k)}$ and $\mathbf{P_0} = Q^n_{\varepsilon,k}$ as a consequence of (5.25) and (5.27) where (5.50) yields (5.49).

Let $\mathbf{P}$ and $\mathbf{Q}$ be probability measures on the space of right-continuous paths with time parameter t running in $[a,b]$. If $\mathbf{Q}$ is Markovian and the marginal distributions at time t of $\mathbf{P}$ and $\mathbf{Q}$ coincide for all $t \in [a,b]$, then $\mathbf{Q}$ is called a Markovian modification of $\mathbf{P}$.

Proposition 5.2.1 *Let $\mathbf{P}$ and $\mathbf{P_0}$ be probability measures on the space of right-continuous (resp. continuous) paths such that $H(\mathbf{P} \mid \mathbf{P_0}) < \infty$. If $\mathbf{P_0}$ is a Markov (resp. diffusion) process, then there exists a Markovian (resp. diffusion) modification $\mathbf{Q}$ of $\mathbf{P}$ with*

$$H(\mathbf{Q} \mid \mathbf{P_0}) \leq H(\mathbf{P} \mid \mathbf{P_0}). \tag{5.50}$$

Proof. Let

$$A = \{\mathbf{R} : \mathbf{R} \circ X_t^{-1} = \mathbf{P} \circ X_t^{-1}, \quad \forall t \in [a,b]\}.$$

For

$$t_j^{(m)} = (a + \frac{j}{2^m}) \wedge b, \quad \text{where } j, m \in I\!N$$

we consider

$$A^{(m)} = \{\mathbf{R} : \mathbf{R} \circ X_{t_j^{(m)}}^{-1} = \mathbf{P} \circ X_{t_j^{(m)}}^{-1} \quad \text{for } j \in I\!N\}, \quad m \in I\!N. \tag{5.51}$$

The sets A and $A^{(m)}$, $m \in I\!N$, are convex and variation closed. Moreover, they satisfy

$$A^{(m)} \supset A^{(m+1)} \supset A \quad \text{and} \quad \bigcap_{m \in I\!N} A^{(m)} = A \tag{5.52}$$

since the paths are assumed to be right-continuous. Thus we receive

$$H(A \mid \mathbf{P}_0) \geq H(A^{(m+1)} \mid \mathbf{P}_0) \geq H(A^{(m)} \mid \mathbf{P}_0). \tag{5.53}$$

Because of $H(\mathbf{P} \mid \mathbf{P}_0) < \infty$, Lemma 5.2.4 provides the Csiszar projections $\mathbf{Q}^{(m)}$ and $\mathbf{Q}$ of $\mathbf{P}_0$ on $A^{(m)}$ and A, respectively. They are uniquely determined by

$$H(\mathbf{Q}^{(m)} \mid \mathbf{P}_0) = H(A^{(m)} \mid \mathbf{P}_0) \quad \text{and} \quad H(\mathbf{Q} \mid \mathbf{P}_0) = H(A \mid \mathbf{P}_0). \tag{5.54}$$

Hence (5.53) yields

$$H(\mathbf{Q}^{(m)} \mid \mathbf{P}_0) \leq H(\mathbf{Q} \mid \mathbf{P}_0) = H(A \mid \mathbf{P}_0) \leq H(\mathbf{P} \mid \mathbf{P}_0) < \infty$$

which implies (5.50). As a consequence of (5.52), (5.53) and (5.54), the lower semi-continuity of the relative entropy implies

$$\lim_{m \nearrow \infty} H(\mathbf{Q}^{(m)} \mid \mathbf{P}_0) = H(\mathbf{Q} \mid \mathbf{P}_0). \tag{5.55}$$

Moreover, Csiszar's inequality (5.19) yields

$$H(\mathbf{Q} \mid \mathbf{P}_0) - H(\mathbf{Q}^{(m)} \mid \mathbf{P}_0) \geq H(\mathbf{Q} \mid \mathbf{Q}^{(m)})$$

because of $\mathbf{Q} \in A^{(m)}$. Hence (5.55) implies

$$\lim_{m \nearrow \infty} H(\mathbf{Q} \mid \mathbf{Q}^{(m)}) = 0 \tag{5.56}$$

i.e., $\mathbf{Q}^{(m)}$ converges to $\mathbf{Q}$ in entropy and hence weakly as m tends to infinity by Lemma 5.2.6.

Proposition 6.2.1 in case of a general Markovian reference process $\mathbf{P}_0$ shows that $\mathbf{Q}^{(m)}$ is a Schrödinger process, and hence Markovian, on each subinterval $[t_j^{(m)}, t_{j+1}^{(m)}]$, $j \in \mathbb{N}$, where it possesses the marginal distributions $\mathbf{P} \circ X_t^{-1}$ for $t = t_j^{(m)}$, $t_{j+1}^{(m)}$, $j \in \mathbb{N}$. Consequently, we are going to deduce that $\mathbf{Q}$ as a particular limit of Markovian processes $\mathbf{Q}^{(m)}$ is Markovian. Let us consider

$$P_r^{(m)} f(X_t) = \mathbf{Q}^{(m)}[f(X_{t+r}) \mid X_t] \quad \text{and} \quad P_r^{(\infty)} f(X_t) = \mathbf{Q}[f(X_{t+r}) \mid X_t]$$

for bounded and measurable functions f where t and $t + r$ are added to the set $\{t_j^{(m)}, j \in \mathbb{N}\}$ if necessary. Then

$$\lim_{m \nearrow \infty} P_r^{(m)} f(X_t) = P_r^{(\infty)} f(X_t), \quad \mathbf{P}\text{-a.s.}$$

and

$$\mathbf{Q}[\prod_{j=1}^{k} g_j(X_{t_j^{(m)}}) f(X_{t_k+r})] = \lim_{m \nearrow \infty} \mathbf{Q}^{(m)}[\prod_{j=1}^{k} g_j(X_{t_j^{(m)}}) f(X_{t_k+r})]$$

$$= \lim_{m \nearrow \infty} \mathbf{Q}^{(m)}[\prod_{j=1}^{k} g_j(X_{t_j^{(m)}}) P_r^{(m)} f(X_{t_k})]$$

$$= \lim_{m \nearrow \infty} \mathbf{P}[\prod_{j=1}^{k} g_j(X_{t_j^{(m)}}) P_r^{(m)} f(X_{t_k})] = \mathbf{P}[\prod_{j=1}^{k} g_j(X_{t_j^{(m)}}) P_r^{(\infty)} f(X_{t_k})]$$

$$= \mathbf{Q}[\prod_{j=1}^{k} g_j(X_{t_j^{(m)}}) P_r^{(\infty)} f(X_{t_k})]$$

for bounded and measurable functions g_j, $j = 1, \ldots, k$. We have used the weak convergence of $(\mathbf{Q}^{(m)})_{m \in \mathbb{N}}$ to $\mathbf{Q}$, the Markov property of $\mathbf{Q}^{(m)}$, the form of the marginal distributions of $\mathbf{Q}^{(m)}$ and the P-a.s. convergence of $(P_r^{(m)})_{m \in \mathbb{N}}$ to $P_r^{(\infty)}$ in precisely this order.

If $\mathbf{P}_0$ is a diffusion process, the *Maruyama-Girsanov drift transformation* in Liptser-Shiryayev (1977)'s chap. 6 and Ikeda-Watanabe (1989)'s chap. 4 can be applied. For simplicity we assume that the diffusion matrix of $\mathbf{P}_0$ equals the identity. Since (5.54) and (5.56) yield $\mathbf{Q} << \mathbf{Q}^{(m)} << \mathbf{P}_0$, there exists a Wiener measure $\mathbf{P}_W$ such that

$$\frac{d\mathbf{Q}}{d\mathbf{P}_W}(X) = \exp\{\int_a^b \mathbf{b}(t, X_t) \, dX_t - \frac{1}{2} \int_a^b | \mathbf{b}(t, X_t) |^2 \, dt\}$$

$$\frac{d\mathbf{Q}^{(m)}}{d\mathbf{P}_W}(X) = \exp\{\int_a^b \mathbf{b}^{(m)}(t, X_t) \, dX_t - \frac{1}{2} \int_a^b | \mathbf{b}^{(m)}(t, X_t) |^2 \, dt\}$$

$\mathbf{Q}$-a.s. and $\mathbf{Q}^{(m)}$-a.s., respectively, where $\mathbf{b}(t, x)$ and $\mathbf{b}^{(m)}(t, x)$ represent the drift coefficients of $\mathbf{Q}$ and $\mathbf{Q}^{(m)}$, respectively. Hence

$$\log(\frac{d\mathbf{Q}}{d\mathbf{Q}^{(m)}})(X) = \int_a^b \mathbf{b}(t, X_t) \, dX_t - \int_a^b \mathbf{b}^{(m)}(t, X_t) \, dX_t$$

$$- \frac{1}{2} \int_a^b | \mathbf{b}(t, X_t) |^2 \, dt + \frac{1}{2} \int_a^b | \mathbf{b}^{(m)}(t, X_t) |^2 \, dt$$

$\mathbf{Q}$-a.s., where its integral with respect to $\mathbf{Q}$ is finite because of (5.56). In terms of the representation $dX_t = dW_t + \mathbf{b}(t, X_t) \, dt$, where $(W_t)_{a \leq t \leq b}$ is the Brownian motion with respect to $\mathbf{Q}$,

$$H(\mathbf{Q} \mid \mathbf{Q}^{(m)}) = \mathbf{Q}[\int_a^b \mathbf{b}(t, X_t) \, dW_t + \int_a^b | \mathbf{b}(t, X_t) |^2 \, dt +$$

$$- \int_a^b \mathbf{b}^{(m)}(t, X_t) \, dW_t - \int_a^b \mathbf{b}^{(m)}(t, X_t) \, \mathbf{b}(t, X_t) \, dt$$

$$- \frac{1}{2} \int_a^b |\mathbf{b}(t, X_t)|^2 \, dt + \frac{1}{2} \int_a^b |\mathbf{b}^{(m)}(t, X_t)|^2 \, dt]$$

$$= \frac{1}{2} \mathbf{Q}[\int_a^b |\mathbf{b}(t, X_t) - \mathbf{b}^{(m)}(t, X_t)|^2 \, dt] \qquad (5.57)$$

which vanishes as m tends to infinity because of (5.56). Hence $\mathbf{b}^{(m)}(t, x)$ converges to $\mathbf{b}(t, x)$ in the L^2-sense of the right-hand side of (5.57). $\Diamond$

Chapter 6

Interacting Diffusion Processes

The phenomenon 'propagation of chaos' is discussed in terms of the relative entropy in (5.26) which allows general microscopical systems and which provides results in the tradition of statistical mechanics. In Section 6.3 we show 'propagation of chaos in entropy' for particle clouds with prescribed initial and final distributions: The particles become asymptotically independent and perform identically according to a Csiszar projection as their number increases to infinity. These limiting distributions turn out to be Schrödinger processes, i.e., diffusion processes considered from Schrödinger (1931)'s time-symmetrical point of view. They are uniquely characterized by the large deviation principle deduced in Chapter 5. The associated rate function minimizes the relative entropy with respect to a renormalized Markovian reference process which has in general singular creation and killing as discussed in Sections 6.2 and 6.4. The n-product of the renormalized reference process conditioned by means of the empirical distribution on an approximation of $A_{a,b}$ in (5.2) possesses a Markovian modification. This system of interacting diffusion processes is proved to perform propagation of chaos in entropy with a Schrödinger process as limiting distribution.

In other words, Schrödinger processes can be considered as Gibbs states realized by microscopical ensembles in terms of systems of interacting diffusion processes. In Chapter 2 we established Schrödinger processes as real-valued counterparts to wave functions. As a consequence, the nature of quantum mechanics can now be explained by Boltzmann's statistical mechanics.

6.1   Eddington-Schrödinger prediction

Schrödinger processes in Definition 2.4.1 are diffusion processes with given dynamics and prescribed initial and final distributions . In Chapter 2 they are constructed by means of transformations in terms multiplicative functionals. The aim of this chapter is to characterize Schrödinger processes by means of a large deviation principle and as limiting distributions of propagation of chaos in entropy. The Markovian reference processes will possess singular creation and killing as it naturally occurs in Chapter 2 with diffusions related to Schrödinger equations. Since the initial and final distributions μ_a and μ_b at $-\infty < a < b < \infty$ are fixed in case of Schrödinger processes, intermediate states at $t \in (a, b)$ will be predicted from past and future. Below we give a note on the conceptual difference between prediction in case of classical deterministic motions and prediction in case of stochastic processes.

When discussing large deviations of Schrödinger processes, we consider the set $A_{a,b}$ in (5.2) of probability measures on the space of continuous paths which have the prescribed pair (μ_a, μ_b) of probability measures on the state space as marginal distributions at time a and b, respectively. Theorem 6.2.1 states the approximate Sanov property of the set $A_{a,b}$ for Schrödinger processes with in general singular creation and killing.

Propagation of chaos in entropy for certain systems of interacting diffusion processes with a Schrödinger process as limiting distribution will be established in Theorem 6.3.1 as a consequence of the approximate Sanov property of the set $A_{a,b}$. A notable implication to quantum and statistical physics will be stated in Corollary 6.3.1. Section 6.2 deals with the renormalization of diffusion processes with creation and killing. Its probabilistical meaning will be clarified by means of a transformation in terms of a multiplicative functional introduced in Section 6.4. Section 6.5 deals with integrability conditions on the creation and killing potential in the situation related to Schrödinger equations.

Note on prediction   In order to realize the significance of Schrödinger processes in quantum physics, we should notice the fact that there are three principal types of prediction:

(I)   prediction of future states based on present data
       (*forward prediction*)

(II)   prediction of past states given final data
       (*backward prediction*)

(III) prediction of intermediate states based on initial and final data
       (*Eddington-Schrödinger prediction*)

In classical deterministical problems the three types lead to the same results. Hence the choice of prediction only depends on convenience. In stochastical problems however, a type of prediction has to be chosen in advance since each one provides a different answer. Actually, the three types of prediction result in *different types of conditioning*, i.e., *different kinds of drift*. This fact has not been well known to physicists. Perhaps only Eddington and Schrödinger realized that the right choice of prediction is crucial for a correct understanding of Schrödinger equations: Quantum mechanics requires Eddington-Schrödinger prediction. This matter of fact is hidden in the complex-valued formulation of quantum mechanics and starts to appear in Schrödinger (1931, 1932)'s investigations on the time-reversibility of natural laws. We motivated Schrödinger's conjecture in Section 1.4, revealed the analytical relations in Section 2.1 and developed a probabilistical framework in Chapters 2 and 5. Corollary 6.3.1 and parts of Section 6.5 will provide applications of Schrödinger processes to quantum mechanics.

6.2   Limiting distributions

Let $(X_t, \ s \leq t \leq b, \ P_{(s,x)}, \ (s,x) \in [a,b] \times I\!\!R^d)$ be the diffusion process in Definition 2.3.1 where $\Omega = C([a,b], I\!\!R^d)$ serves as the actual path space. Let c be a measurable extended real-valued function on $[a,b] \times I\!\!R^d$ with positive and negative parts denoted by $c^+ = c \vee 0$ and $c^- = -c \vee 0$, respectively. We intend to consider space-time diffusion processes with in general singular creation and killing $c = c^+ - c^-$. In terms of the killing part c^-, we set

$$T_s = \inf\{t > s : \int_s^t c^-(v, X_v)\, dv = \infty\} \tag{6.1}$$

and determine the relevant state space $D \subset [a, b] \times \mathbb{R}^d$ as

$$D = \{(s, x) : P_{(s,x)}(b < T_s) > 0\}. \tag{6.2}$$

The measure $P_{(s,x)}^-$ with killing c^- can now be defined as

$$P_{(s,x)}^-[F] = P_{(s,x)}[\exp\{-\int_s^b c^-(v, X_v)\, dv\}\, F\, 1_{\{b < T_s\}}] \tag{6.3}$$

for non-negative measurable functions F on Ω. We notice that the exponential function on the right-hand side of (6.3) is positive on the set $\{b < T_s\}$ because of (6.1). Hence $P_{(s,x)}^-[1] > 0$ for all $(s, x) \in D$, i.e., the probability that a particle starting from a point in D survives until the final time b is positive.

Dealing with the creation part c^+, we establish the basic integrability condition

$$\xi(s, x) = P_{(s,x)}^-[\exp\{\int_s^b c^+(v, X_v)\, dv\}] < \infty \tag{6.4}$$

for all $(s, x) \in D$, where $P_{(s,x)}^-$ is defined in (6.3) and $\xi(s, x) > 0$ holds for all $(s, x) \in D$ because of (6.2). We are now prepared to treat the process $P_{(s,x)}^c$ with creation and killing c given by

$$P_{(s,x)}^c[F] = P_{(s,x)}[\exp\{\int_s^b c(v, X_v)\, dv\}\, F\, 1_{\{b < T_s\}}] \tag{6.5}$$

for all $(s, x) \in D$. In terms of the measure $P_{(s,x)}^c$, the function ξ on D introduced in (6.4) is nothing but

$$\xi(s, x) = P_{(s,x)}^c[1] \quad \text{for all} \quad (s, x) \in D.$$

Under the integrability assumption (6.4), the *renormalized process* $\overline{P}$ of $P_{(s,x)}^c$ can be defined as

$$\overline{P}[F] = \int_{D_a} d\varsigma(x)\, \frac{1}{\xi(a, x)}\, P_{(a,x)}^c[F] \tag{6.6}$$

where $d\varsigma(x) = \varsigma(x)\, dx$ with $\varsigma > 0$ on $D_a = \{x : (a, x) \in D\}$ is any fixed initial distribution. The renormalized process (6.6) is going to play the role of the Markovian reference process $\overline{P}$ in Section 5.1. Its properties will be investigated in Section 6.4.

In order to involve the two observed, i.e., prescribed, marginal distributions considered in Section 1.1, we introduce the set $\mathcal{E}(a, b)$ of all probability measures on $(I\!R^d \times I\!R^d, \sigma(I\!R^d \times I\!R^d))$ with the prescribed pair (μ_a, μ_b) of probability measures on $I\!R^d$ as their marginal distributions. Particle dynamics from initial time a directly to final time b can be represented by a probability measure $\bar{p}$ on $I\!R^d \times I\!R^d$ which is determined by

$$\bar{p}(A \times B) = \int_{D_a} d\varsigma(x) \frac{1}{\xi(a,x)} 1_A(x) P^c_{(a,x)}[1_B(X_b)] \qquad (6.7)$$

for $A, B \in \sigma(I\!R^d)$. The assumption

$$\exists \, p \in \mathcal{E}(a,b) \; : \; H(p \mid \bar{p}) < \infty \qquad (6.8)$$

imposes a relation between particle dynamics and marginal constraints, where the definition of the relative entropy $H(\,.\mid .\,)$ is given in (5.4). Condition (6.8) implies $p(\,.\times I\!R^d) << \bar{p}(\,.\times I\!R^d)$; hence the distribution μ_a must be absolutely continuous w.r.t. the distribution ς. Moreover, $p(I\!R^d \times .\,) << \bar{p}(I\!R^d \times .\,)$ yields $\bar{p}(I\!R^d \times \mathrm{supp}\,\mu_b) > 0$ which means that the process in (6.5) with creation and killing c must reach the support of the distribution μ_b at the final time b with positive probability. Now we can refer to

Lemma 6.2.1 (Csiszar 1975, Föllmer 1988, Nagasawa 1990)
If (6.4) and (6.8) are provided, then there exists the up to multiplicative constants unique non-negative solution $(\hat{\varphi}, \varphi)$ of the Schrödinger system

$$\int_{D_a} \hat{\varphi}(x) \, 1_A(x) \, dx \, P^c_{(a,x)}[\varphi(X_b)] \;\; = \;\; \mu_a(A)$$

$$\int_{D_a} \hat{\varphi}(x) \, dx \, P^c_{(a,x)}[\varphi(X_b) \, 1_B(X_b)] \;\; = \;\; \mu_b(B)$$

for all $A, B \in \sigma(I\!R^d)$ where $P^c_{(a,x)}$ is provided by (6.5).

Equivalently, there exists a unique solution $q(dx\,dz) \in \mathcal{E}(a,b)$ of the variational problem

$$H(q \mid \bar{p}) = \inf_{p \in \mathcal{E}(a,b)} H(p \mid \bar{p}). \qquad (6.9)$$

The solutions $(\hat{\varphi}, \varphi)$ and $q(dx\,dz)$ satisfy the relation

$$\frac{dq}{d\bar{p}}(x,z) = \frac{\xi(a,x)}{\varsigma(x)} \hat{\varphi}(x)\,\varphi(z) \qquad (6.10)$$

for μ_a-a.a. $x \in I\!R^d$ and μ_b-a.a. $z \in I\!R^d$ with the properties $\log \hat{\varphi} \in L^1(\mu_a)$ and $\log \varphi \in L^1(\mu_b)$.

Lemma 6.2.1 actually claims that $q(dx\,dz)$ is the unique Csiszar projection in Lemma 5.2.4 of $\overline{p}(dx\,dz)$ in (6.7) on the set $\mathcal{E}(a, b)$. Since $\mathcal{E}(a, b)$ is convex and variation closed, Lemma 6.2.1 is an immediate consequence of Csiszar (1975)'s theorem 3.1,

In (6.10) we find a factorization in products of functions in x and z, respectively, which is deduced in Csiszar (1975) by means of exponential families. Multiplicative constants $k > 0$ and $1/k$ can obviously be chosen freely for $\hat{\varphi}$ and φ, respectively, and even differently in regions separated by their zero sets. However, they will not have any influence in the crucial expression (6.11) below. A Lagrangian approach in Section 1.3 yielded the product form (1.16) which is (6.10) for discrete situations. Chapter 7 will provide an approach in terms of a product measure endomorphism which will again yield the form of (6.10), but deduced under different assumptions.

Lemma 6.2.2 *Let (6.4) and (6.8) be provided. Then Csiszar's projection Q of the renormalized process $\overline{P}$ in (6.6) on the set $A_{a,b}$ in (5.2) can be represented uniquely in terms of solutions $(\hat{\varphi}, \varphi)$ of the Schrödinger system in Lemma 6.2.1 and the process $P^c_{(a,x)}$ in (6.5) with creation and killing c as*

$$Q[F] = \int_{D_a} \hat{\varphi}(x)\,dx\,P^c_{(a,x)}[F\,\varphi(X_b)] \tag{6.11}$$

for bounded measurable functions F on Ω where (5.3) and (6.8) turn out to be equivalent.

Proof. Let

$$P^z_x = P(.\mid X_a = x, X_b = z)$$

and let

$$p(dx\,dz) = P(X_a \in dx,\ X_b \in dz)$$

for any $P \in M_1(\Omega)$. Hence, if $P \in M_1(\Omega)$ such that $P << \overline{P}$ on $\sigma(\Omega)$, then

$$\frac{dP}{d\overline{P}} = \frac{dP^z_x}{d\overline{P}^z_x}\frac{dp}{d\overline{p}}.$$

Lemma 5.2.4 establishes Csiszar's projection as the minimum point of the variational principle (5.18). We notice that $P \in A_{a,b}$ if and only if the associated p

belongs to $\mathcal{E}(a,b)$. The condition (5.3) provides at least one $P \in A_{a,b}$ such that

$$
\begin{aligned}
H(P \mid \overline{P}) &= \int_{\Omega} \log(\frac{dP}{d\overline{P}}) \, dP \\
&= \int_{\mathbb{R}^d \times \mathbb{R}^d \times \Omega} \log(\frac{dP_x^z}{d\overline{P}_x^z}) \, dP_x^z \, p(dx \, dz) + \int_{\mathbb{R}^d \times \mathbb{R}^d \times \Omega} \log(\frac{dp}{d\overline{p}}) \, dP_x^z \, p(dx \, dz) \\
&= \int_{\mathbb{R}^d \times \mathbb{R}^d} H(P_x^z \mid \overline{P}_x^z) \, p(dx \, dz) + H(p \mid \overline{p}) < \infty.
\end{aligned}
$$

Hence the assumptions (5.3) and (6.8) as well as the variational problems (5.18) and (6.9) are equivalent since the relative entropy is non-negative. Lemma 6.2.1 provides a solution of (6.9) in terms of the process Q in (6.11). Thus Q also solves (5.18), i.e., it is Csiszar's projection of $\overline{P}$ on $A_{a,b}$ as a consequence of the uniqueness stated in Lemma 5.2.4. $\Diamond$

Proposition 6.2.1 *Provided (6.4) and (6.8), $(X_t, \, a \le t \le b, \, Q)$ in (6.11) is a Schrödinger process of Definition 2.4.1 for the triplet (c, μ_a, μ_b).*

Proof. Following Lemma 6.2.2, the process Q in (6.11) is a Markov field in $M_1(\Omega)$. It possesses the representation required by Definition 2.4.1 because of Theorem 2.3.1. Hence, only the Markov property of Q remains to be verified.

We notice that analogously to (3.6) and (6.50), the space-time function

$$
\begin{aligned}
\varphi(t,x) &= P_{(t,x)}[\exp\{\int_t^b c(v, X_v) \, dv\} \, \varphi(X_b) \, 1_{\{b < T_t\}}] \quad &(6.12) \\
&= P_{(t,x)}^-[\exp\{\int_t^b c^+(v, X_v) \, dv\} \, \varphi(X_b)]
\end{aligned}
$$

in terms of $P_{(t,x)}^-$ in (6.3) is a solution of the integral equation

$$
\varphi(t,x) = P_{(t,x)}^-[\varphi(X_b)] + \int_t^b du \, P_{(t,x)}^-[c^+(u, X_u) \, \varphi(u, X_u)]
$$

for all $(t,x) \in D$. Hence, $\varphi(t,x)$ is an excessive function of $P_{(t,x)}^-$ which means that $\varphi(t, X_t)$ is right-continuous in $t \in [s,b]$ as a consequence of Blumenthal-Getoor (1968)'s theorem 4.8. The Markov property of $P_{(s,x)}$ yields

$$
Q[F] = \int_{D_a} \hat{\varphi}(x) \, dx \, P_{(a,x)}[\exp\{\int_a^t c(v, X_v) \, dv\} \, F \, \varphi(t, X_t) \, 1_{\{t < T_a\}}]
$$

for $t \in [a, b]$ and bounded, measurable space-time functions F. Thus the right-continuity of $\varphi(t, X_t)$ implies

$$Q((t, X_t)_{a \leq t \leq b} \text{ does not hit the zero set of } \varphi) = 1. \qquad (6.13)$$

In order to deduce the Markov property of Q, let G be any non-negative, $\sigma((v, X_v), \; a \leq v \leq s)$-measurable function and let f be any non-negative, measurable function on $[a, b] \times \mathbb{R}^d$. Then the Markov property of $P_{(s,x)}$ claims that

$$Q[G \, f(t, X_t)]$$
$$= \int_{D_a} \hat{\varphi}(x) \, dx \, P_{(a,x)}[\exp\{\int_a^s c(v, X_v) \, dv\} \, G \, \varphi(s, X_s) \, q(s, X_s; t, f) \, 1_{\{s < T_a\}}]$$
$$= Q[G \, q(s, X_s; t, f)]$$

for $s < t$ and $D_a = \{x \in \mathbb{R}^d : (a, x) \in D\}$ where the transition probability

$$q(s, x; t, f) \qquad (6.14)$$
$$= P_{(s,x)}[\varphi(s, X_s)^{-1} \, \exp\{\int_s^t c(v, X_v) \, dv\} \, \varphi(t, X_t) \, f(t, X_t) \, 1_{\{t < T_s\}}]$$

is well-defined because of (6.13).

Alternatively, the Markov property of a Schrödinger process Q can be shown by means of N_s^t, $a \leq s < t \leq b$, in Definition 2.3.2 which is well-defined because of (6.13). In fact, the Markov property of $P_{(s,x)}$ provides

$$\varphi(s, x) = P_{(s,x)}[\exp\{\int_s^t c(v, X_v) \, dv\} \, \varphi(t, X_t) \, 1_{\{t < T_s\}}] \qquad (6.15)$$

for $(s, x) \in D$ which means $P_{(s,x)}[N_s^t] = 1$. Hence Corollary 2.3.2 yields the Markovian representation

$$Q[F] = \int_{D_a} \hat{\varphi}(x) \, \varphi(a, x) \, P_{(a,x)}[N_a^b \, F]$$

which has the transition probability $q(s, x; t, f)$ found in (6.14). $\diamond$

As a consequence of Proposition 6.2.1, Theorem 5.1.1 yields

Theorem 6.2.1 *Provided (6.4) and (6.8), there exist the renormalized process $\overline{P}$ in (6.6) and the Schrödinger process Q in (6.11). They allow in the notation of Section 5.1 to claim that:*

(i) The set $A_{a,b}$ in (5.2) possesses the approximate Sanov property

$$\lim_{k \nearrow \infty} \lim_{n \nearrow \infty} \frac{1}{n} \log \overline{\mathbf{P}}(L_n \in A(\varepsilon, k)) = -H(Q \mid \overline{P})$$

where $A(\varepsilon, k) \searrow A_{a,b}$ as $k \nearrow \infty$. In other words, the probability that the rare event $\{L_n \in A(\varepsilon, k)\}$ occurs is given by

$$\overline{\mathbf{P}}(L_n \in A(\varepsilon, k)) \approx e^{-n\,H(Q|\overline{P})} \quad \text{for } n \nearrow \infty \text{ and } k \nearrow \infty.$$

(ii) The system $\overline{\mathbf{P}}^{(n,k)}$ in (5.9) is weakly asymptotically quasi-independent *with Csiszar's projection Q as limiting distribution, i.e.,*

$$\lim_{k \nearrow \infty} \lim_{n \nearrow \infty} \frac{1}{n} H(Q^n(\,.\, \mid L_n \in A(\varepsilon, k)) \mid \overline{\mathbf{P}}^{(n,k)}) = 0 \tag{6.16}$$

where Q satisfies

$$\lim_{n \nearrow \infty} Q^n(L_n \in A(\varepsilon, k)) = 1, \quad \forall k \in I\!N.$$

6.3 Propagation of chaos in entropy

Propagation of chaos for systems of interacting diffusion processes is introduced by McKean (1966, 1967) and investigated in Gutkin-Kac (1983), Tanaka (1984), Kusuoka-Tamura (1984), Ölschläger (1989), Sznitman (1989) and Dawson-Gärtner (1989). Our discussion is related to problems considered by Nagasawa-Tanaka (1986, 1987a,b). We formulate propagation of chaos in a comparably strong form, appropriate for Schrödinger processes. Convergence will be described by means of relative entropy which has been established as a natural indicator for the mutual randomness of probability measures. See for example Csiszar (1975, 1984) in connection with large deviations, Boltzmann (1896) and Lanford (1973) who deal with statistical mechanics and Khinchin (1957) for an approach to information theory . Different types of convergence will be compared in Remark 6.3.3 where convergence in entropy will turn out to be the strongest one.

Definition 6.3.1 A sequence of systems $(\,(X_1, \ldots, X_n),\ \mathbf{Q}^{(n,k)}\,)$ of interacting diffusion processes performs *propagation of chaos in entropy* with limiting distribution Q as n and k tend to infinity if for each $m \in I\!N$, the marginal distribution $\mathbf{Q}_m^{(n,k)}$ of $\mathbf{Q}^{(n,k)}$ on Ω^m and the empirical distribution L_n in (5.5) satisfy

$$\lim_{k \nearrow \infty} \lim_{n \nearrow \infty} H(Q^m \mid \mathbf{Q}_m^{(n,k)}) = 0 \qquad (6.17)$$

$$\lim_{k \nearrow \infty} \lim_{n \nearrow \infty} H(Q \mid \mathbf{Q}^{(n,k)}[L_n]) = 0 \qquad (6.18)$$

where k is called modeling parameter.

Definition 6.3.1 allows general microscopical systems represented by $\mathbf{Q}^{(n,k)}$ as they are required in the situation of Section 1.1 for generalized dynamics in (6.6). In fact, while $Q^m << \mathbf{Q}_m^{(n,k)}$ seems to be quite reasonable, the equivalence of measures does restrict the choice of microscopical systems too much. As a consequence, our limiting distribution Q appears as the first argument in the non-symmetrical relative entropy formula $H(\,.\,\mid\,.\,)$ in (5.4). This causes the main technical difficulties we have to deal with in the proof of Theorem 6.3.1.

Remark 6.3.1 Propagation of chaos in entropy is introduced in Aebi-Nagasawa (1992a). Adapted to the non-symmetrical relative entropy formula in (5.4), Nagasawa (1993)'s refined definition consists of a limiting procedure in two steps. First, $H(\mathbf{Q}_m^{(n,k)} \mid Q_k^m)$ as well as $H(\mathbf{Q}_m^{(n,k)}[L_n] \mid Q_k)$ are required to vanish as n tends to infinity where $(Q_k)_{k \in I\!\!N}$ is any family of approximative limiting distributions indexed by the modeling parameter k. Second, $H(Q \mid Q_k)$ is required to vanish as k tends to infinity. Our Definition 6.3.1 comes back to a direct convergence in entropy. Thus Theorem 6.3.1 will claim propagation of chaos following the tradition of statistical mechanics in Schrödinger (1931)'s situation of Section 1.1.

Let us reveal the meaning of Definition 6.3.1. In one sentence we can say that (6.17) guarantees the existence of a 'typical particle' and (6.18) ensures that asymptotically it is the 'typical particle' which is observed. More precisely, (6.17) describes m arbitrary of the interacting diffusion processes in terms of their joint m-dimensional distribution $\mathbf{Q}_m^{(n,k)}$ which converges in entropy to Q^m. This means that any m interacting diffusion processes are asymptotically independent and that Q is the limiting distribution of each interacting diffusion process. Property (6.18) deals with a mixing of the involved interacting diffusion processes by means of L_n in (5.5). It represents the observed distribution on Ω under $\mathbf{Q}^{(n,k)}$ which tends in entropy to Q as the number n of participating diffusion processes and the modeling parameter k increase to infinity.

Our intention is to furnish Schrödinger processes Q with systems of interacting diffusion processes $\mathbf{Q}^{(n,k)}$ which perform propagation of chaos in the sense of Definition 6.3.1. In fact, we claim

Theorem 6.3.1 *If (6.4) and (6.8) are provided, then:*

(i)   There exists a Markovian modification $\mathbf{Q}^{(n,k)}$ of $\overline{\mathbf{P}}^{(n,k)}$ in (5.9) on Ω^n. ($\mathbf{X}_t = (X_t^{(1)}, \ldots, X_t^{(n)})$, $a \leq t \leq b$, $\mathbf{Q}^{(n,k)}$) represents a system of interacting diffusion processes with the Markovian drift coefficient $\mathbf{b}^{(n,k)}(t, \mathbf{x})$, $\mathbf{x} = (x^{(1)}, \ldots, x^{(n)}) \in (I\!\!R^d)^n$, which describes the interaction in terms of

$$\mathbf{b}^{(n,k)}(t, \mathbf{x}) = (\, \mathbf{b}_i^{(n,k)}(t, \mathbf{x}, L_n(\mathbf{x}))\,)_{i=1,\ldots,n} \tag{6.19}$$

where $\mathbf{b}_i^{(n,k)}$ is the drift vector with respect to the process $(X_t^{(i)}$, $a \leq t \leq b$, $Q_{\varepsilon,k})$ in (5.26).

(ii)   The Markovian modification $\mathbf{Q}^{(n,k)}$ is weakly asymptotically quasi-independent with the Schrödinger process Q in (6.11) as limiting distribution, i.e.,

$$\lim_{k \nearrow \infty} \lim_{n \nearrow \infty} \frac{1}{n} H(Q^n(\,.\,\mid L_n \in A(\varepsilon, k)) \mid \mathbf{Q}^{(n,k)}) = 0 \tag{6.20}$$

where

$$\lim_{n \nearrow \infty} Q^n(L_n \in A(\varepsilon, k)) = 1.$$

(iii)   The system ($\mathbf{X}_t$, $a \leq t \leq b$, $\mathbf{Q}^{(n,k)}$) performs 'propagation of chaos in entropy' of Definition 6.3.1, at least along subsequences, with Csiszar's projection Q in (5.41) as limiting distribution when n and k tend to infinity, provided $\overline{P}^m$-a.s. convergent subsequences of $\log(\mathbf{Q}_m^{(n,k)}/Q_{\varepsilon,k}^m)$, n, $k \in I\!\!N$, with $Q^m << \mathbf{Q}_m^{(n,k)}$ are uniformly integrable w.r.t. Q^m for $m \in I\!\!N$. An alternative assumption is that for every $m \in I\!\!N$, subsequences $(n(\nu), k(\nu))$, $\nu \in I\!\!N$, with $Q^m << \mathbf{Q}_{m,\cdot}^{(n(\nu),k(\nu))}$, along which $\log(\mathbf{Q}_m^{(n(\nu),k(\nu))}/Q_{\varepsilon,k(\nu)}^m)$ is $\overline{P}^m$-a.s. convergent, satisfy $\forall \delta > 0$, $\exists M < \infty$, $\exists \delta_1 > 0$, $\exists \nu(m, \delta, M, \delta_1) \in I\!\!N$ such that

$$\int_{B_{\nu,M}^{(m)}} \left(\frac{d\mathbf{Q}_m^{(n(\nu),k(\nu))}}{dQ_{\varepsilon,k(\nu)}^m} \right)^t dQ_{\varepsilon,k(\nu)}^m < \delta \tag{6.21}$$

for all $t \in (-\delta_1, \delta_1)$ and for all $\nu > \nu(m, \delta, M, \delta_1)$ where

$$B_{\nu,M}^{(m)} = \{(\frac{dQ_m^{(n(\nu),k(\nu))}}{dQ_{\varepsilon,k(\nu)}^m})^\theta > M \quad for \ \ \theta = 1 \ or \ -1\}. \tag{6.22}$$

The integrability condition (6.21) shows the required accuracy of the approximation $A(\varepsilon, k)$ in (5.7) of $A_{a,b}$ in (5.2) with respect to the reference measure $\overline{P}$.

In view of verifying propagation of chaos claimed in Theorem 6.3.1, we are going to deduce some properties of finite-dimensional marginal distributions.

Lemma 6.3.1 *Let $\mathbf{P}_m^{(n)}$ be the marginal distribution of $\mathbf{P}^{(n)} \in M_1(\Omega^n)$, $n \in \mathbb{N}$, on $\Omega^m = \Omega_1 \times \ldots \times \Omega_m$ for $m \leq n$. Let $P \in M_1(\Omega)$ such that $H(\mathbf{P}^{(n)} \mid P^n) < \infty$ for $n \in \mathbb{N}$. If the marginal distributions of $\mathbf{P}^{(n)}$ on $\Omega_{m\nu+1} \times \ldots \times \Omega_{m(\nu+1)}$ for $\nu = 1, 2, \ldots$ equal $\mathbf{P}_m^{(n)}$, then*

$$H(\mathbf{P}_m^{(n)} \mid P^m) \leq \frac{m}{n-r} H(\mathbf{P}^{(n)} \mid P^n) \tag{6.23}$$

where r is a non-negative integer smaller than m.

Proof. By assumption $\mathbf{P}^{(n)}$ possesses identical marginal distributions $\mathbf{P}_m^{(n)}$ on $\Omega_1 \times \ldots \times \Omega_m$ and on $\Omega_{m+1} \times \ldots \times \Omega_{2m}$ etc. For $m \in \mathbb{N}$, let $s \in \mathbb{N}$ be so large that $n = ms + r$ with $0 \leq r < m$. Then the coinciding marginal distributions imply

$$\mathbf{P}^{(n)} << (\mathbf{P}_m^{(n)})^s \mathbf{P}_r^{(n)} << P^n.$$

Since the relative entropy is non-negative,

$$H(\mathbf{P}^{(n)} \mid P^n) = \int \log(\frac{d\mathbf{P}^{(n)}}{dP^n}) \, d\mathbf{P}^{(n)}$$

$$= \int \log(\frac{d\mathbf{P}^{(n)}}{(d\mathbf{P}_m^{(n)})^s \cdot d\mathbf{P}_r^{(n)}}) \, d\mathbf{P}^{(n)} + \int \log(\frac{(d\mathbf{P}_m^{(n)})^s \cdot d\mathbf{P}_r^{(n)}}{dP^n}) \, d\mathbf{P}^{(n)}$$

$$\geq \int \log(\frac{(d\mathbf{P}_m^{(n)})^s \cdot d\mathbf{P}_r^{(n)}}{dP^n}) \, d\mathbf{P}^{(n)}$$

$$= s \int \log(\frac{d\mathbf{P}_m^{(n)}}{dP^m}) \, d\mathbf{P}_m^{(n)} + \int \log(\frac{d\mathbf{P}_r^{(n)}}{dP^r}) \, d\mathbf{P}_r^{(n)}$$

$$\geq s \int \log(\frac{d\mathbf{P}_m^{(n)}}{dP^m}) \, d\mathbf{P}_m^{(n)}$$

which is (6.23) because of $s = (n-r)/m$. $\Diamond$

We denote a partial empirical distribution by

$$L_{m,n}(\mathbf{X}) = \frac{1}{n-m} \sum_{j=m+1}^{n} \delta_{X^{(j)}}, \quad m < n, \quad \mathbf{X} = (X^{(1)}, \ldots, X^{(n)}) \in \Omega^n.$$

(6.24)

Lemma 6.3.2 *Let us assume (5.3). (i) If $n > m\, 2^k/\varepsilon$ is chosen large enough such that*

$$\frac{n-m}{n} L_{m,n}(\mathbf{X}) \in A(\varepsilon - \tilde{\varepsilon}(\frac{m}{n}), k)$$

where

$$\tilde{\varepsilon}(\frac{m}{n}) = 2^k \frac{m}{n}$$

then

$$L_n(\mathbf{X}) \in A(\varepsilon, k)$$

for $\overline{P}^n$-a.a. $\mathbf{X} \in \Omega^n$.

(ii) Moreover, there exists $n(m, \varepsilon, k) \in \mathbb{N}$, $n(m, \varepsilon, k) > m\, 2^k/\varepsilon$, such that

$$Q^m << \overline{\mathbf{P}}_m^{(n,k)}$$

(6.25)

for all $n \geq n(m, \varepsilon, k)$ where $\overline{\mathbf{P}}_m^{(n,k)}$ is the m-dimensional marginal distribution of $\overline{\mathbf{P}}^{(n,k)}$ in (5.9).

Proof. Let $\mathcal{P}_k = \{B_1, \ldots, B_k\}$ be a partition of $\mathbb{R}^d$ as introduced in (5.6). In view of the definition of $A(\varepsilon, k)$ in (5.7), we first notice for $r = a, b$ that $L_n(\mathbf{X}) \circ X_r^{-1}$ is not absolutely continuous with respect to $\overline{P} \circ X_r^{-1}$ on the set $\{\mathbf{X} \in \Omega^n : \exists j = 1, \ldots, n,\ \exists i = 1, \ldots, k$ such that $X^{(j)} \in X_r^{-1}(B_i)$ where $\overline{P}(X_r^{-1}(B_i)) = 0\}$ which is however a $\overline{P}^n$-zero set. In terms of (6.24) we receive

$$| L_n(\mathbf{X})(X_r^{-1}(B_i)) - q_r(B_i) |$$

$$\leq | \frac{1}{n} \sum_{j=1}^{m} \delta_{X^{(j)}}(X_r^{-1}(B_i)) | + | \frac{1}{n} \sum_{j=m+1}^{n} \delta_{x^{(j)}}(X_r^{-1}(B_i)) - q_r(B_i) |$$

$$= | \frac{m}{n} L_m(\mathbf{X})(X_r^{-1}(B_i)) | + | \frac{n-m}{n} L_{m,n}(\mathbf{X})(X_r^{-1}(B_i)) - q_r(B_i) |$$

for $i = 1, \ldots, k$ and $r = a, b$. Hence (i) follows for $\overline{P}^n$-almost every $\mathbf{X} \in \Omega^n$, if n is so large that

$$\frac{\varepsilon - \tilde{\varepsilon}(\frac{m}{n})}{2^k} = \frac{\varepsilon}{2^k} - \frac{m}{n}$$

is positive.

Let $n > m\, 2^k/\varepsilon$ and let $B^{(m)} \in \sigma(\Omega^m)$. Because of $Q << \overline{P}$ in (ii) of Lemma 5.2.9, part (i) implies

$$Q^n(B^{(m)} \cap \{L_n \in A(\varepsilon, k)\}) \tag{6.26}$$

$$\geq \quad Q^n(B^{(m)} \cap \{\frac{n-m}{n} L_{m,n} \in A(\varepsilon - \tilde\varepsilon(\frac{m}{n}), k)\})$$

$$= \quad Q^m(B^{(m)})\, Q^n(\frac{n-m}{n} L_{m,n} \in A(\varepsilon - \tilde\varepsilon(\frac{m}{n}), k)),$$

Since the law of large numbers claims that

$$Q^n(\frac{n-m}{n} L_{m,n} \in A(\varepsilon - \tilde\varepsilon(\frac{m}{n}), k)) \to 1 \quad \text{as } n \nearrow \infty$$

there exists $n(m, \varepsilon, k) > m\, 2^k/\varepsilon$ such that for all $n \geq n(m, \varepsilon, k)$

$$Q^n(\frac{n-m}{n} L_{m,n} \in A(\varepsilon - \tilde\varepsilon(\frac{m}{n}), k)) > 0.$$

If $Q^m(B^{(m)}) > 0$ then inequality (6.26) yields

$$Q^n(B^{(m)} \cap \{L_n \in A(\varepsilon, k)\}) > 0$$

for all $n \geq n(m, \varepsilon, k)$ where $n(m, \varepsilon, k)$ is independent of the particular set $B^{(m)}$. As a consequence of $Q << \overline{P}$,

$$\overline{P}^n(B^{(m)} \cap \{L_n \in A(\varepsilon, k)\}) > 0$$

which implies

$$\overline{\mathbf{P}}_m^{(n,k)}(B^{(m)}) = \overline{\mathbf{P}}^{(n,k)}(B^{(m)}) > 0$$

concluding the proof. $\Diamond$

Lemma 6.3.3 *Let (5.3) be provided and let* $\overline{\mathbf{P}}_m^{(n,k)}$ *be the m-dimensional marginal distribution of the conditional process* $\overline{\mathbf{P}}^{(n,k)}$ *in (5.9). If* $n > m\, 2^k/\varepsilon$ *then*

$$\tilde p_{m,n,k}\, d\overline{P}^m \leq d\overline{\mathbf{P}}_m^{(n,k)} \leq p_{m,n,k}\, d\overline{P}^m$$

where $0 < \tilde p_{m,n,k} \leq 1 \leq p_{m,n,k} < \infty$ *are real numbers for every* $m,\ k \in \mathbb{N}$ *and* $\varepsilon > 0$.

Proof. Let $n > m \, 2^k / \varepsilon$ be large enough such that Lemma 6.3.2 holds. Then

$$d\overline{\mathbf{P}}_m^{(n,k)} = \int_{(X^{(m+1)},\ldots,X^{(n)})\in\Omega^{n-m}} d\overline{\mathbf{P}}^{(n,k)}(X^{(1)},\ldots,X^{(n)})$$

$$= \frac{1}{\overline{P}^n(L_n \in A(\varepsilon,k))} \int_{(X^{(m+1)},\ldots,X^{(n)})\in\Omega^{n-m}}$$
$$d\overline{P}^n((X^{(1)},\ldots,X^{(m)},X^{(m+1)},\ldots,X^{(n)}) \in \{L_n \in A(\varepsilon,k)\})$$

$$\geq \frac{1}{\overline{P}^n(L_n \in A(\varepsilon,k))} \int_{(X^{(m+1)},\ldots,X^{(n)})\in\Omega^{n-m}}$$
$$d\overline{P}^n((X^{(1)},\ldots,X^{(n)}) \in \{\frac{n-m}{n} L_{m,n} \in A(\varepsilon - \tilde{\varepsilon}(\frac{m}{n}),k)\})$$

$$= \tilde{p}_{m,n,k} \, d\overline{P}^m, \quad \overline{P}^m\text{-a.s.}$$

where

$$0 < \tilde{p}_{m,n,k} = \frac{\overline{P}^{n-m}(\frac{n-m}{n} L_{m,n} \in A(\varepsilon - \tilde{\varepsilon}(\frac{m}{n}),k))}{\overline{P}^n(L_n \in A(\varepsilon,k))} \leq 1.$$

In view of the estimate from above, we notice that (5.11) yields $k(\varepsilon) \in \mathbb{N}$ such that $\overline{P} \notin A(\varepsilon,k)$ for all $k \geq k(\varepsilon)$ since $\overline{P} \notin A_{a,b}$. Hence it may be assumed that $\{L_n \in A(\varepsilon,k)\}$ is monotonously decreasing under $\overline{P}^n$ as n tends to infinity for each $k \geq k(\varepsilon)$. Consequently,

$$d\overline{P}^n((X^{(1)},\ldots,X^{(n)}) \in \{L_n \in A(\varepsilon,k)\}) \leq$$

$$\leq d\overline{P}^n((X^{(1)},\ldots,X^{(m)},\ldots,X^{(n)}) \in \{\frac{1}{n-m} \sum_{j=m+1}^{n} \delta_{X^{(j)}} \in A(\varepsilon,k)\})$$

$$= d\overline{P}^m \, d\overline{P}^{n-m}((X^{(m+1)},\ldots,X^{(n)}) \in \{L_{m,n} \in A(\varepsilon,k)\})$$

which yields

$$d\overline{\mathbf{P}}_m^{(n,k)} = \frac{1}{\overline{P}^n(L_n \in A(\varepsilon,k))} \int_{(X^{(m+1)},\ldots,X^{(n)})\in\Omega^{n-m}}$$
$$d\overline{P}^n((X^{(1)},\ldots,X^{(m)},X^{(m+1)},\ldots,X^{(n)}) \in \{L_n \in A(\varepsilon,k)\})$$

$$\leq p_{m,n,k} \, d\overline{P}^m, \quad \overline{P}^m\text{-a.s.}$$

for

$$1 \leq p_{m,n,k} = \frac{\overline{P}^{n-m}(L_{m,n} \in A(\varepsilon,k))}{\overline{P}^n(L_n \in A(\varepsilon,k))} < \infty.$$

Lemma 6.3.4 *Let us collect some frequently used results.*

(*i*) *Let* a_n, b_n, $n \in \mathbb{N}$, *be two sequences such that* $-\infty < \liminf_{n \nearrow \infty} z_n$ *and* $\limsup_{n \nearrow \infty} z_n < \infty$ *for* $z_n = b_n$ *and* $z_n = a_n + b_n$, $n \in \mathbb{N}$. *Then*

$$\liminf_{n \nearrow \infty}(a_n + b_n) - \limsup_{n \nearrow \infty} b_n \leq \liminf_{n \nearrow \infty} a_n \leq \limsup_{n \nearrow \infty} a_n$$
$$\leq \limsup_{n \nearrow \infty}(a_n + b_n) - \liminf_{n \nearrow \infty} b_n.$$

(*ii*) *Let* $(S, \mathcal{F}, \mu)$ *be a measurable space and let* f, f_n, $n \in \mathbb{N}$, *be measurable,* μ-*a.e. finite functions on* S *such that*

$$\lim_{n \nearrow \infty} \int_S \mid f_n(x) - f(x) \mid d\mu(x) = 0.$$

Then there exists a subsequence $n(\nu)$, $\nu \in \mathbb{N}$, *such that*

$$\lim_{\nu \nearrow \infty} f_{n(\nu)}(x) = f(x) \quad for \ \mu\text{-}a.a. \ x \in S.$$

Moreover, if there exists a subsequence $n(\nu_1)$, $\nu_1 \in \mathbb{N}$, *and a measurable,* μ-*a.e. finite function* $f^{(1)}$ *on* S *such that*

$$\lim_{\nu_1 \nearrow \infty} f_{n(\nu_1)}(x) = f^{(1)}(x) \quad for \ \mu\text{-}a.a. \ x \in S$$

then

$$f = f^{(1)}, \quad \mu\text{-}a.e. \ on \ S.$$

(*iii*) $x \log x \geq e^{-1}$, $\forall x \in [0, \infty)$.

Proof. (*i*) easily follows by means of

$$\limsup_{n \nearrow \infty} a_n = \limsup_{n \nearrow \infty}(a_n + b_n - b_n) \leq \limsup_{n \nearrow \infty}(a_n + b_n) + \limsup_{n \nearrow \infty}(-b_n)$$
$$= \limsup_{n \nearrow \infty}(a_n + b_n) - \liminf_{n \nearrow \infty} b_n$$

where

$$\liminf_{n \nearrow \infty} a_n = -\limsup_{n \nearrow \infty}(-a_n).$$

Concerning (*ii*) we refer to Kingman-Taylor (1977)'s theorem 7.2 and notice that

$$\liminf_{\nu_1 \nearrow \infty} \int_S \mid f_{n(\nu_1)}(x) - f(x) \mid d\mu(x) \geq \int_S \mid f^{(1)}(x) - f(x) \mid d\mu(x) \geq 0$$

by Fatou's lemma. To check (*iii*) is routine. $\diamondsuit$

Lemma 6.3.5 *Let us assume (5.3). Then the limit*

$$\lim_{k \nearrow \infty} H(Q \mid Q_{\varepsilon,k}) = 0$$

in (5.48) of Csiszar projections Q and $Q_{\varepsilon,k}$ in (iii) of Lemma 5.2.9 implies convergence in variation, i.e.,

$$\lim_{k \nearrow \infty} \int \mid \frac{dQ_{\varepsilon,k}}{d\overline{P}} - \frac{dQ}{d\overline{P}} \mid d\overline{P} = 0. \tag{6.27}$$

Moreover,

$$\liminf_{k \nearrow \infty} \frac{dQ_{\varepsilon,k}}{d\overline{P}} = \limsup_{k \nearrow \infty} \frac{dQ_{\varepsilon,k}}{d\overline{P}} = \frac{dQ}{d\overline{P}}, \quad \overline{P}\text{-}a.s. \tag{6.28}$$

which is $\overline{P}$-a.s. finite and

$$\liminf_{k \nearrow \infty} \frac{dQ}{dQ_{\varepsilon,k}} = \limsup_{k \nearrow \infty} \frac{dQ}{dQ_{\varepsilon,k}} = 1, \quad Q\text{-}a.s. \tag{6.29}$$

Finally, if $B_k \in \sigma(\Omega)$, $k \in I\!N$, is such that $\liminf_{k \nearrow \infty} 1_{B_k} = \limsup_{k \nearrow \infty} 1_{B_k}$, Q-a.s., then

$$\liminf_{k \nearrow \infty} \int_{B_k} \log(\frac{dQ}{dQ_{\varepsilon,k}})\, dQ = \limsup_{k \nearrow \infty} \int_{B_k} \log(\frac{dQ}{dQ_{\varepsilon,k}})\, dQ = 0. \tag{6.30}$$

Proof. Because of (5.20), (5.48) provides (6.27). Hence Lemma 6.3.4 (*ii*) yields (6.28) as well as (6.29) because of (5.47) since $dQ_{\varepsilon,k}/d\overline{P}$ and $dQ/d\overline{P}$ are $\overline{P}$-a.s. finite as a consequence of (5.41), (5.42) and assumption (5.3).

Following its definition in (5.6),

$$H(Q \mid Q_{\varepsilon,k}) = \int \log(\frac{dQ}{dQ_{\varepsilon,k}}) \frac{dQ}{dQ_{\varepsilon,k}} \frac{dQ_{\varepsilon,k}}{d\overline{P}}\, d\overline{P}$$

where

$$1_{B_k} (\log(\frac{dQ}{dQ_{\varepsilon,k}}) \frac{dQ}{dQ_{\varepsilon,k}} + e^{-1}) \frac{dQ_{\varepsilon,k}}{d\overline{P}} \geq 0$$

by Lemma 6.3.4 (*iii*). Hence Fatou's lemma yields

$$\liminf_{k \nearrow \infty} \int_{B_k} (\log(\frac{dQ}{dQ_{\varepsilon,k}}) \frac{dQ}{dQ_{\varepsilon,k}} + e^{-1}) \frac{dQ_{\varepsilon,k}}{d\overline{P}}\, d\overline{P}$$

$$\geq \int \liminf_{k \nearrow \infty} 1_{B_k} [\liminf_{k \nearrow \infty} (\log(\frac{dQ}{dQ_{\varepsilon,k}}) \frac{dQ}{dQ_{\varepsilon,k}}) + e^{-1}] \liminf_{k \nearrow \infty} (\frac{dQ_{\varepsilon,k}}{d\overline{P}})\, d\overline{P}$$

$$= e^{-1} Q[\liminf_{k \nearrow \infty} 1_{B_k}]$$

where (6.28) and (6.29) have been employed. Analogously for $\Omega \backslash B_k$ instead of B_k,

$$
\begin{aligned}
&\liminf_{k \nearrow \infty} \int 1_{\Omega \backslash B_k} \, (\log(\frac{dQ}{dQ_{\varepsilon,k}}) \frac{dQ}{dQ_{\varepsilon,k}} + e^{-1}) \frac{dQ_{\varepsilon,k}}{d\overline{P}} \, d\overline{P} \\
&\geq \quad e^{-1} \, Q[\liminf_{k \nearrow \infty}(1 - 1_{B_k})] = e^{-1} \, (1 - Q[\limsup_{k \nearrow \infty} 1_{B_k}]).
\end{aligned}
$$

Consequently,

$$
\begin{aligned}
&\limsup_{k \nearrow \infty} \int 1_{B_k} \, (\log(\frac{dQ}{dQ_{\varepsilon,k}}) \frac{dQ}{dQ_{\varepsilon,k}} + e^{-1}) \frac{dQ_{\varepsilon,k}}{d\overline{P}} \, d\overline{P} \\
&= \quad \lim_{k \nearrow \infty} H(Q \mid Q_{\varepsilon,k}) + e^{-1} \\
&\quad - \liminf_{k \nearrow \infty} \int 1_{\Omega \backslash B_k} \, (\log(\frac{dQ}{dQ_{\varepsilon,k}}) \frac{dQ}{dQ_{\varepsilon,k}} + e^{-1}) \frac{dQ_{\varepsilon,k}}{d\overline{P}} \, d\overline{P} \\
&\leq \quad e^{-1} \, Q[\limsup_{k \nearrow \infty} 1_{B_k}].
\end{aligned}
$$

Hence Lemma 6.3.4 (i) applied in case of

$$
a_k = \int 1_{B_k} \, \log(\frac{dQ}{dQ_{\varepsilon,k}}) \, dQ \quad \text{and} \quad b_k = e^{-1} \, Q_{\varepsilon,k}[1_{B_k}] \quad \text{for} \quad k \in I\!\!N
$$

yields

$$
\begin{aligned}
&e^{-1}(Q[\liminf_{k \nearrow \infty} 1_{B_k}] - \limsup_{k \nearrow \infty} Q_{\varepsilon,k}[1_{B_k}]) \leq \liminf_{k \nearrow \infty} \int 1_{B_k} \log(\frac{dQ}{dQ_{\varepsilon,k}}) \, dQ \\
&\leq \quad \limsup_{k \nearrow \infty} \int 1_{B_k} \log(\frac{dQ}{dQ_{\varepsilon,k}}) \, dQ \leq e^{-1} \, (Q[\limsup_{k \nearrow \infty} 1_{B_k}] - \liminf_{k \nearrow \infty} Q_{\varepsilon,k}[1_{B_k}]).
\end{aligned}
$$

The proof is completed by Fatou's lemma which claims

$$
\begin{aligned}
&\limsup_{k \nearrow \infty} Q_{\varepsilon,k}[1_{B_k}] = 1 - \liminf_{k \nearrow \infty} \int 1_{\Omega \backslash B_k} \frac{dQ_{\varepsilon,k}}{dQ} \, dQ \\
&\leq \quad 1 - \int \liminf_{k \nearrow \infty} 1_{\Omega \backslash B_k} \, dQ = Q[\limsup_{k \nearrow \infty} 1_{B_k}]
\end{aligned}
$$

as well as

$$
\liminf_{k \nearrow \infty} Q_{\varepsilon,k}[1_{B_k}] \geq \int \liminf_{k \nearrow \infty} 1_{B_k} \, \liminf_{k \nearrow \infty} (\frac{dQ_{\varepsilon,k}}{dQ}) \, dQ = Q[\liminf_{k \nearrow \infty} 1_{B_k}]
$$

because of (6.29). $\Diamond$

Proposition 6.3.1 *Let us assume condition (5.3) and let $Q_{\varepsilon,k}$ be the process in (ii) of Lemma 5.2.9. Then, there exists a Markovian modification $\mathbf{Q}^{(n,k)}$ of $\overline{\mathbf{P}}^{(n,k)}$ in (5.9) with property (5.49). Its m-dimensional marginal distributions $\mathbf{Q}_m^{(n,k)}$, $m \in \mathbb{N}$, satisfy*

$$\lim_{k \nearrow \infty} \lim_{n \nearrow \infty} H(\mathbf{Q}_m^{(n,k)} \mid Q_{\varepsilon,k}^m) = 0 \qquad (6.31)$$

and hence

$$\lim_{k \nearrow \infty} \lim_{n \nearrow \infty} \int \left| \frac{dQ_{\varepsilon,k}^m}{d\overline{P}^m} - \frac{d\mathbf{Q}_m^{(n,k)}}{d\overline{P}^m} \right| d\overline{P}^m = 0. \qquad (6.32)$$

Moreover,

$$\liminf_{k \nearrow \infty} \liminf_{n \nearrow \infty} \frac{d\mathbf{Q}_m^{(n,k)}}{dQ_{\varepsilon,k}^m} = \limsup_{k \nearrow \infty} \limsup_{n \nearrow \infty} \frac{d\mathbf{Q}_m^{(n,k)}}{dQ_{\varepsilon,k}^m} = 1, \quad Q^m\text{-}a.s. \qquad (6.33)$$

If $(n(\nu), k(\nu))$, $\nu \in \mathbb{N}$, is any subsequence along which

$$g_\nu^{(m)} = \frac{d\mathbf{Q}_m^{(n(\nu),k(\nu))}}{dQ_{\varepsilon,k(\nu)}^m} \qquad (6.34)$$

is $\overline{P}^m$-a.s. convergent, then

$$\liminf_{\nu \nearrow \infty} \int_{\tilde{B}_{\nu,M}^{(m)}} \log(g_\nu^{(m)}) \, d\mathbf{Q}_m^{(n(\nu),k(\nu))} = \limsup_{\nu \nearrow \infty} \int_{\tilde{B}_{\nu,M}^{(m)}} \log(g_\nu^{(m)}) \, d\mathbf{Q}_m^{(n(\nu),k(\nu))} = 0 \qquad (6.35)$$

with

$$\tilde{B}_{\nu,M}^{(m)} = \{ |\log(g_\nu^{(m)})| \geq M \} \qquad (6.36)$$

for every constant $M > 0$.

Proof. The existence of $\mathbf{Q}^{(n,k)}$ with (5.49) is provided by Proposition 6.3.1 on the basis of (5.14) and (5.42). In view of (6.31) we notice that the marginal distributions of $\overline{\mathbf{P}}^{(n,k)}$ and of $\mathbf{Q}^{(n,k)}$ on $\Omega_{m\nu+1} \times \ldots \times \Omega_{m(\nu+1)}$ for $\nu = 0, 1, 2, \ldots$ coincide obviously because of the symmetry of $\overline{\mathbf{P}}^{(n,k)}$ in $X^{(i)}$, $i = 1, \ldots, n$. Hence

$$H(\mathbf{Q}_m^{(n,k)} \mid Q_{\varepsilon,k}^m) \leq m \frac{n}{n-r} \frac{1}{n} H(\mathbf{Q}^{(n,k)} \mid Q_{\varepsilon,k}^n) \qquad (6.37)$$

by Lemma 6.3.1 where $0 \leq r \leq m - 1$. As a consequence of (5.49), the right-hand side of (6.37) vanishes as n and k tend to infinity because of (5.14). (6.32) follows from (6.31) by means of (5.20).

For a proof of (6.35), we first verify that the sequence $d(n, k)$, $n, k \in \mathbb{N}$, defined by

$$d(2n, k) = \frac{d\mathbf{Q}_m^{(n,k)}}{d\overline{P}^m} \quad \text{and} \quad d(2n + 1, k) = \frac{dQ_{\varepsilon,k}^m}{d\overline{P}^m}$$

is a Cauchy sequence in variation distance, i.e., $\forall\, \delta > 0$, $\exists k_0$, $n_0 \in \mathbb{N}$ such that

$$\int |\, d(n_1, k_1) - d(n_2, k_2)\,|\; d\overline{P}^m < \delta \qquad (6.38)$$

$\forall\, k_1$, $k_2 > k_0$ and $\forall\, n_1$, $n_2 > n_0$. By means of (5.20),

$$\int |\, d(2n, k) - d(2n + 1, k)\,|\; d\overline{P}^m \le \sqrt{2H(\mathbf{Q}_m^{(n,k)}\,|\,Q_{\varepsilon,k}^m)} \qquad (6.39)$$

which vanishes as n and k tend to infinity because of (6.31). In (6.38) there are three possible combinations; the first one is

$$
\begin{aligned}
|\,\mathbf{Q}_m^{(n_1,k_1)} - \mathbf{Q}_m^{(n_2,k_2)}\,|_{var} \;=\;& |\,\mathbf{Q}_m^{(n_1,k_1)} - Q_{\varepsilon,k_1}^m\,|_{var} \\
&+\; |\,Q_{\varepsilon,k_1}^m - Q_{\varepsilon,k_2}^m\,|_{var} \;+\; |\,Q_{\varepsilon,k_2}^m - \mathbf{Q}_m^{(n_2,k_2)}\,|_{var}
\end{aligned}
$$

where the first and third member of the right-hand side vanish in (6.39). The second member which is the second combination

$$|\,Q_{\varepsilon,k_1}^m - Q_{\varepsilon,k_2}^m\,|_{var} \le |\,Q_{\varepsilon,k_1}^m - Q^m\,|_{var} + |\,Q^m - Q_{\varepsilon,k_2}^m\,|_{var}$$

vanishes as k tends to infinity because of (5.48). It remains the third combination

$$|\,\mathbf{Q}_m^{(n_1,k_1)} - Q_{\varepsilon,k_2}^m\,|_{var} \le |\,\mathbf{Q}_m^{(n_1,k_1)} - Q_{\varepsilon,k_1}^m\,|_{var} + |\,Q_{\varepsilon,k_1}^m - Q_{\varepsilon,k_2}^m\,|_{var}$$

which also vanishes in (6.39).

Since $(M_1(\Omega^m)$, $|\cdot|_{var})$ is complete, the Cauchy sequence of probability measures with densities $d(n, k)$, n, $k \in \mathbb{N}$, converges in variation to some probability distribution $\mathbf{P}_1 \in M_1(\Omega^m)$, i.e.,

$$\lim_{k \nearrow \infty}\lim_{n \nearrow \infty} \int |\, d(n, k) - \frac{d\mathbf{P}_1}{d\overline{P}^m}\,|\; d\overline{P}^m = 0.$$

Let $(k(\nu_1), n(\nu_1))$, $\nu_1 \in \mathbb{N}$, be any sequence such that $\lim_{\nu_1 \nearrow \infty} d(k(\nu_1), n(\nu_1))$ exists, $\overline{P}^m$-a.s. According to Lemma 6.3.4 (ii), there is at least one such sequence and as a subsequence of a Cauchy sequence it satisfies

$$\lim_{\nu_1 \nearrow \infty} \int |\, d(n(\nu_1), k(\nu_1)) - \frac{d\mathbf{P}_1}{d\overline{P}^m}\,|\; d\overline{P}^m = 0.$$

Thus we receive

$$\lim_{\nu_1 \nearrow \infty} d(k(\nu_1), n(\nu_1)) = \frac{d\mathbf{P}_1}{d\overline{P}^m}, \quad \overline{P}^m\text{-a.s.}$$

by Fatou's lemma.

The densities $d(2n, k)$ and $d(2n + 1, k)$, n, $k \in I\!\!N$, represent two subsequences of the Cauchy sequence $d(n, k)$, hence they are convergent in variation distance with the same limit. Moreover, there exists a subsequence along which $d\mathbf{Q}_m^{(n,k)}/d\overline{P}^m$ converges $\overline{P}^m$-a.s. to $d\mathbf{P}_1/d\overline{P}^m$ and there exists a subsequence $k(\nu_2)$, $\nu_2 \in I\!\!N$, such that

$$\lim_{\nu_2 \nearrow \infty} \frac{dQ_{\varepsilon,k(\nu_2)}^m}{d\overline{P}^m} = \frac{d\mathbf{P}_1}{d\overline{P}^m}, \quad \overline{P}^m\text{-a.s.}$$

Fatou's lemma yields

$$Q^m = \mathbf{P}_1, \quad Q^m\text{-a.s.}$$

since

$$\int \left| \frac{dQ^m}{d\overline{P}^m} - \frac{d\mathbf{P}_1}{d\overline{P}^m} \right| d\overline{P}^m = \int \lim_{\nu_2 \nearrow \infty} \left| \frac{dQ^m}{d\overline{P}^m} - \frac{dQ_{\varepsilon,k(\nu_2)}^m}{d\overline{P}^m} \right| d\overline{P}^m$$

$$\leq \liminf_{\nu_2 \nearrow \infty} \int \left| \frac{dQ^m}{d\overline{P}^m} - \frac{dQ_{\varepsilon,k(\nu_2)}^m}{d\overline{P}^m} \right| d\overline{P}^m \leq \sqrt{2m} \lim_{\nu_2 \nearrow \infty} \sqrt{H(Q \mid Q_{\varepsilon,k(\nu_2)})} = 0$$

where (5.20) and (5.48) have been employed.

Let $(n(\nu), k(\nu))$, $\nu \in I\!\!N$, be any subsequence along which $g_\nu^{(m)}$ in (6.34) is $\overline{P}^m$-a.s. convergent. Because of (6.28) there exists a subsequence $(n(\nu_3), k(\nu_3))$, $\nu_3 \in I\!\!N$, of $(n(\nu), k(\nu))$, $\nu \in I\!\!N$, such that

$$\lim_{\nu_3 \nearrow \infty} \frac{dQ_{\varepsilon,k(\nu_3)}}{d\overline{P}} = \frac{dQ}{d\overline{P}}, \quad \overline{P}\text{-a.s.}$$

Because of

$$\int \left| \lim_{\nu_3 \nearrow \infty} g_{\nu_3}^{(m)} - \lim_{\nu_3 \nearrow \infty} \frac{dQ^m}{dQ_{\varepsilon,k(\nu_3)}^m} \right| \lim_{\nu_3 \nearrow \infty} \left(\frac{dQ_{\varepsilon,k(\nu_3)}^m}{d\overline{P}^m} \right) d\overline{P}^m$$

$$= \int \lim_{\nu_3 \nearrow \infty} \left| \frac{d\mathbf{Q}_m^{(n(\nu_3),k(\nu_3))}}{dQ_{\varepsilon,k(\nu_3)}^m} - \frac{dQ^m}{dQ_{\varepsilon,k(\nu_3)}^m} \right| \frac{dQ_{\varepsilon,k(\nu_3)}^m}{d\overline{P}^m} d\overline{P}^m$$

$$\leq \liminf_{\nu_3 \nearrow \infty} \int \left| \frac{d\mathbf{Q}_m^{(n(\nu_3),k(\nu_3))}}{d\overline{P}^m} - \frac{dQ^m}{d\overline{P}^m} \right| d\overline{P}^m = 0$$

provided by Fatou's lemma, property (6.29) yields

$$\lim_{\nu \nearrow \infty} g_\nu^{(m)} = \lim_{\nu_3 \nearrow \infty} g_{\nu_3}^{(m)} = 1, \quad Q^m\text{-a.s.}$$

which implies (6.33).

As a consequence of (6.33), $\tilde{B}_{\nu,M}^{(m)}$ in (6.36) satisfies

$$\liminf_{\nu \nearrow \infty} 1_{\tilde{B}_{\nu,M}^{(m)}} = \limsup_{\nu \nearrow \infty} 1_{\tilde{B}_{\nu,M}^{(m)}} = 0, \quad \overline{P}^m\text{-a.s.}$$

for any $M > 0$ and for any subsequence $(n(\nu), k(\nu))$, $\nu \in \mathbb{N}$, providing $\overline{P}^m$-a.s. convergence in (6.34). Moreover,

$$\int_B \log(g_\nu^{(m)}) \, d\mathbf{Q}_m^{(n(\nu),k(\nu))} + e^{-1} Q_{\varepsilon,k(\nu)}^m(B) \qquad (6.40)$$

$$= \int_B [\log(g_\nu^{(m)}) \, g_\nu^{(m)} + e^{-1}] \frac{dQ_{\varepsilon,k(\nu)}^m}{d\overline{P}^m} \, d\overline{P}^m$$

for any $B \in \sigma(\Omega)$ where the integrand on the right-hand side of (6.40) is non-negative according to Lemma 6.3.4 (*iii*). Hence Fatou's lemma yields

$$\liminf_{\nu \nearrow \infty} \int 1_{\tilde{B}_{\nu,M}^{(m)}} [\log(g_\nu^{(m)}) \, g_\nu^{(m)} + e^{-1}] \frac{dQ_{\varepsilon,k(\nu)}^m}{d\overline{P}^m} \, d\overline{P}^m$$

$$\geq \int \liminf_{\nu \nearrow \infty} (1_{\tilde{B}_{\nu,M}^{(m)}} [\log(g_\nu^{(m)}) \, g_\nu^{(m)} + e^{-1}] \frac{dQ_{\varepsilon,k(\nu)}^m}{d\overline{P}^m}) \, d\overline{P}^m$$

$$= e^{-1} Q^m[\liminf_{\nu \nearrow \infty} 1_{\tilde{B}_{\nu,M}^{(m)}}] = 0$$

where (6.28) and (6.33) have been employed. There is the analogous relation for $\Omega \backslash \tilde{B}_{\nu,M}^{(m)}$ in place of $\tilde{B}_{\nu,M}^{(m)}$. Thus

$$\limsup_{\nu \nearrow \infty} \int 1_{\tilde{B}_{\nu,M}^{(m)}} [\log(g_\nu^{(m)}) \, g_\nu^{(m)} + e^{-1}] \frac{dQ_{\varepsilon,k(\nu)}^m}{d\overline{P}^m} \, d\overline{P}^m$$

$$= \lim_{\nu \nearrow \infty} \int \log(g_\nu^{(m)}) \, d\mathbf{Q}_m^{(n(\nu),k(\nu))} + e^{-1}$$

$$- \liminf_{\nu \nearrow \infty} \int 1_{\Omega \backslash \tilde{B}_{\nu,M}^{(m)}} [\log(g_\nu^{(m)}) \, g_\nu^{(m)} + e^{-1}] \frac{dQ_{\varepsilon,k(\nu)}^m}{d\overline{P}^m} \, d\overline{P}^m$$

$$\leq e^{-1}(1 - Q^m[\liminf_{\nu \nearrow \infty} 1_{\Omega \backslash \tilde{B}_{\nu,M}^{(m)}}]) = e^{-1} Q^m[\limsup_{\nu \nearrow \infty} 1_{\tilde{B}_{\nu,M}^{(m)}}] = 0$$

where (6.31) has been used. Because of (6.40) we are now in the situation of Lemma 6.3.4 (*i*) which claims that (6.35) follows, if

$$\liminf_{\nu \nearrow \infty} Q_{\varepsilon,k(\nu)}^m[1_{\tilde{B}_{\nu,M}^{(m)}}] = \limsup_{\nu \nearrow \infty} Q_{\varepsilon,k(\nu)}^m[1_{\tilde{B}_{\nu,M}^{(m)}}].$$

In fact,

$$| Q^m_{\varepsilon,k(\nu)}[1_{\tilde{B}^{(m)}_{\nu,M}}] | \leq | Q^m[1_{\tilde{B}^{(m)}_{\nu,M}}] | + | Q^m_{\varepsilon,k(\nu)}[1_{\tilde{B}^{(m)}_{\nu,M}}] - Q^m[1_{\tilde{B}^{(m)}_{\nu,M}}] |$$

$$\leq | Q^m[1_{\tilde{B}^{(m)}_{\nu,M}}] | + \int 1_{\tilde{B}^{(m)}_{\nu,M}} | \frac{dQ^m}{d\overline{P}^m} - \frac{dQ^m_{\varepsilon,k(\nu)}}{d\overline{P}^m} | \, d\overline{P}^m$$

where the right-hand side vanishes as ν tends to infinity for any subsequence in (6.34) because of Fatou's lemma in case of the first member and because of (5.20) and (5.48) in case of the second member. $\diamond$

Lemma 6.3.6 *Let (5.3) and (6.21) be provided. If for $m \in \mathbb{N}$, $g^{(m)}_\nu$, $\nu \in \mathbb{N}$, is given in (6.34), then $\forall \delta > 0$, $\exists M < \infty$, $\exists \nu(\delta, M) \in \mathbb{N}$ such that*

$$\int_{\tilde{B}^{(m)}_{\nu,M}} | \log(g^{(m)}_\nu) | \, dQ^m < \delta, \quad \forall \nu > \nu(\delta, M) \tag{6.41}$$

with $\tilde{B}^{(m)}_{\nu,M}$ in (6.36).

Proof. Referring to (6.34) we notice that for $M \in (0, \infty)$

$$\tilde{B}^{(m)}_{\nu,\tilde{M}} = \{| \log g^{(m)}_\nu | > \tilde{M}\} = \{(g^{(m)}_\nu)^\theta > e^{\tilde{M}} \quad \text{for} \quad \theta = 1 \text{ or } -1\}$$

corresponds to $B^{(m)}_{\nu,M}$ in (6.22) with $M = e^{\tilde{M}}$. If, as provided by (6.21),

$$\int_{B^{(m)}_{\nu,M}} \exp\{t \log(g^{(m)}_\nu)\} \, dQ^m_{\varepsilon,k(\nu)}$$

$$= \int_{B^{(m)}_{\nu,M} \cap \{(\log(g^{(m)}_\nu))^+ > 0\}} \exp\{t \, (\log(g^{(m)}_\nu))^+\} \, dQ^m_{\varepsilon,k(\nu)}$$

$$+ \int_{B^{(m)}_{\nu,M} \cap \{(\log(g^{(m)}_\nu))^- > 0\}} \exp\{-t \, (\log(g^{(m)}_\nu))^-\} \, dQ^m_{\varepsilon,k(\nu)} < \frac{\tilde{\delta}}{2}$$

for all $t \in (-\delta_1, \delta_1)$ where $1 < M < \infty$, then

$$\int_{B^{(m)}_{\nu,M}} \exp\{t \, | \log(g^{(m)}_\nu) |\} \, dQ^m_{\varepsilon,k(\nu)} < \tilde{\delta} \tag{6.42}$$

for all $t \in (-\delta_1, \delta_1)$. In order to arrive at (6.41), the inequality

$$ab < a \log a + e^b \quad \text{for} \quad a, b \geq 0$$

in Beckenbach-Bellmann (1961)'s section 15 is applied to $a_\nu = dQ^m/dQ^m_{\varepsilon,k(\nu)}$ and $b_\nu = t \mid \log(g_\nu^{(m)}) \mid$ for $t \in (0,\delta_1)$ and $\nu \in I\!N$. Thus

$$\int_{B^{(m)}_{\nu,M}} \mid \log(g_\nu^{(m)}) \mid dQ^m = \frac{1}{t} \int_{B^{(m)}_{\nu,M}} \frac{dQ^m}{dQ^{(m)}_{\varepsilon,k(\nu)}} t \mid \log(g_\nu^{(m)}) \mid dQ^m_{\varepsilon,k(\nu)}$$

$$< \frac{1}{t} \int_{B^{(m)}_{\nu,M}} \frac{dQ^m}{dQ^{(m)}_{\varepsilon,k(\nu)}} \log(\frac{dQ^m}{dQ^{(m)}_{\varepsilon,k(\nu)}}) \, dQ^m_{\varepsilon,k(\nu)}$$

$$+ \frac{1}{t} \int_{B^{(m)}_{\nu,M}} \exp\{t \mid \log(g_\nu^{(m)}) \mid\} \, dQ^m_{\varepsilon,k(\nu)} < \delta$$

since the first member of the right-hand side vanishes as k tends to infinity as a consequence of (6.30) for $(n(\nu), k(\nu))$, $\nu \in I\!N$, in (6.34), while the second member is smaller than δ according to the proof of Lemma 5.2.6 and (6.42).$\diamondsuit$

Remark 6.3.2 In order to verify propagation of chaos, i.e., (6.17) and (6.18), we essentially have for use the limit (5.48) which is a consequence of (5.42) and (5.47) and the limit (6.31) which is a consequence of (5.14), (5.49) and (6.37). While (6.31) provides some integrability properties in Proposition 6.3.1, Lemma 6.3.6 shows some consequences of (5.48) under assumption (6.21). Let us briefly resume our approach. Because of (5.42) and (5.48), $Q << Q_{\varepsilon,k} << \overline{P}$ where $Q_{\varepsilon,k}$ is Csiszar's projection of $\overline{P}$ on $A(\varepsilon,k)$ in Lemma 5.2.9 (ii). Following Lemma 6.3.2, $Q^m << \mathbf{Q}^{(n,k)}_m$ for $n > n(m,\varepsilon,k)$ where $\mathbf{Q}^{(n,k)}$ is the Markovian modification of $\overline{\mathbf{P}}^{(n,k)}$ in (5.9) provided by Proposition 6.3.1. Hence Lemma 6.3.3 yields

$$\frac{1}{p_{m,n,k}} \frac{dQ^m}{d\overline{P}^m} \leq \frac{dQ^m}{d\mathbf{Q}^{(n,k)}_m} \leq \frac{1}{\tilde{p}_{m,n,k}} \frac{dQ^m}{d\overline{P}^m}, \quad \overline{P}^m\text{-a.s.}$$

for $n > n(m,\varepsilon,k)$. Thus

$$-\log p_{m,n,k} + m\, H(Q \mid \overline{P}) \leq H(Q^m \mid \mathbf{Q}^{(n,k)}_m) \leq -\log \tilde{p}_{m,n,k} + H(Q \mid \overline{P}) \quad (6.43)$$

for $n > n(m,\varepsilon,k)$, which is however inappropriate for a limit consideration as n and k tend to infinity. We notice that (6.27) and (6.32) yield

$$\lim_{k \nearrow \infty} \lim_{n \nearrow \infty} \int \mid \frac{d\mathbf{Q}^{(n,k)}_m}{d\overline{P}^m} - \frac{dQ^m}{d\overline{P}^m} \mid d\overline{P}^m = 0$$

i.e., convergence in variation. To receive $\lim_{k \nearrow \infty} \lim_{n \nearrow \infty} H(Q^m \mid \mathbf{Q}^{(n,k)}_m) = 0$, i.e., (6.17), at least along subsequences, $\mathbf{Q}^{(n,k)}_m$ has to satisfy additionally a uniform integrability condition with respect to the limiting distribution Q^m. Lemma 6.3.3, Lemma 6.3.6 and Proposition 6.3.1 refine (6.27) and (6.32) on the basis of (5.48) and (6.31), respectively. All that remains to be assumed is (6.21) which is a condition rather on the approximating sequence $Q_{\varepsilon,k}$ than on Q itself.

Proof of Theorem 6.3.1. (i) claims that

$$X_t^{(i)} = X_a^{(i)} + \int_a^t \sigma^{(n,k)}(v, \mathbf{X}_v)\, dB_v^{(i)} + \int_a^t \mathbf{b}_i^{(n,k)}(v, \mathbf{X}_v)\, dv \qquad (6.44)$$

for $\mathbf{X}_t = (X_t^{(1)}, \ldots, X_t^{(n)})$ where $\sigma^{(n,k)}$ are uniformly elliptic diffusion coefficients and $(B_t^{(i)})_{a \leq t \leq b}$, $i = 1, \ldots, n$, are independent d-dimensional Brownian motions with respect to $Q_{\varepsilon,k}$. In fact, the Markovian modification $\mathbf{Q}^{(n,k)}$ in Proposition 6.3.1 is obtained as a piecewise Schrödinger process. Moreover, Lemma 5.2.7 and Lemma 5.2.9 imply $H(\mathbf{Q}^{(n,k)} \mid Q_{\varepsilon,k}^n) < \infty$ as well as $H(Q_{\varepsilon,k} \mid \overline{P}) < \infty$ for $n, k \in \mathbb{N}$ because of (5.49) where $Q_{\varepsilon,k}$ is Csiszar's projection of $\overline{P}$ on $A(\varepsilon, k)$. Thus the generalized version of the Maruyama-Girsanov drift transformation considered in the proof of Theorem 2.3.1 yields (6.44) and consequently (6.19).

The weak asymptotical quasi-independence of $\mathbf{Q}^{(n,k)}$ with limiting distribution Q in (6.11) claimed in (ii) follows from (6.16) by means of Proposition 6.3.1.

When starting to prove (iii), we notice that our situation based on (5.9) is particularly symmetrical in $X^{(i)}$, $i = 1, \ldots, n$. Thus, property (6.17) for $m = 1$ implies property (6.18) in Definition 6.3.1.

In order to show (6.17) for any $m \in \mathbb{N}$, we notice first that $H(Q^m \mid \mathbf{Q}_m^{(n,k)})$ is well-defined for $n \geq n(m, \varepsilon, k)$ and $k \in \mathbb{N}$ as a consequence of Lemma 6.3.2, Lemma 6.3.3 and Proposition 6.3.1 shown in (6.43). Since $\mathbf{Q}_m^{(n,k)}$ takes different positions in the relative entropy expressions in (6.17) and (6.31), respectively, we consider

$$\frac{dQ^m}{d\mathbf{Q}_m^{(n,k)}} = \frac{dQ^m}{dQ_{\varepsilon,k}^m} \frac{dQ_{\varepsilon,k}^m}{d\mathbf{Q}_m^{(n,k)}}, \qquad Q^m\text{-a.s.} \qquad (6.45)$$

for $n \geq n(m, \varepsilon, k)$ where the first factor on the right-hand side of (6.45) exists because of (5.48) and the second factor on the right-hand side of (6.45) exists on supp $\mathbf{Q}_m^{(n,k)}$ because of (6.31). As a consequence of (6.45),

$$H(Q^m \mid \mathbf{Q}_m^{(n,k)}) = m\, H(Q \mid Q_{\varepsilon,k}) - \int \log(g_\nu^{(m)})\, dQ^m \qquad (6.46)$$

where $g_\nu^{(m)}$, $\nu \in I\!N$, is defined in (6.34).

Let us investigate the limit behavior of the second member of the right-hand side of (6.46). Proposition 6.3.1 yields because of (6.33) a subsequence $(n(\nu), k(\nu))$, $\nu \in I\!N$, with $n(\nu) > n(m, \varepsilon, k(\nu))$ determined in Lemma 6.3.2 (ii) such that $\lim_{\nu \nearrow \infty} g_\nu^{(m)} = 1$, $\overline{P}^m$-a.s. As a consequence of (6.35), $g_\nu^{(m)}$, $\nu \in I\!N$, may be supposed to satisfy

$$\lim_{\nu \nearrow \infty} \int_{\{|\log(g_\nu^{(m)})|>M\}} \log(g_\nu^{(m)})\, d\mathbf{Q}_m^{(n(\nu),k(\nu))} = 0 \qquad (6.47)$$

for any $M > 0$. Now we are ready to claim that

$$\left| \int \log(g_\nu^{(m)})\, dQ^m - \int \log(g_\nu^{(m)})\, d\mathbf{Q}_m^{(n(\nu),k(\nu))} \right| \qquad (6.48)$$

$$\leq \quad \left| \int \log(g_\nu^{(m)})\, dQ^m - \int_{\{|\log(g_\nu^{(m)})|\leq M\}} \log(g_\nu^{(m)})\, dQ^m \right|$$

$$+ \quad \left| \int_{\{|\log(g_\nu^{(m)})|\leq M\}} \log(g_\nu^{(m)})\, dQ^m \right.$$

$$\left. - \int_{\{|\log(g_\nu^{(m)})|\leq M\}} \log(g_\nu^{(m)})\, dQ_{\varepsilon,k(\nu)}^m \right|$$

$$+ \quad \left| \int_{\{|\log(g_\nu^{(m)})|\leq M\}} \log(g_\nu^{(m)})\, dQ_{\varepsilon,k(\nu)}^m \right.$$

$$\left. - \int_{\{|\log(g_\nu^{(m)})|\leq M\}} \log(g_\nu^{(m)})\, d\mathbf{Q}_m^{(n(\nu),k(\nu))} \right|$$

$$+ \quad \left| \int_{\{|\log(g_\nu^{(m)})|\leq M\}} \log(g_\nu^{(m)})\, d\mathbf{Q}_m^{(n(\nu),k(\nu))} - \int \log(g_\nu^{(m)})\, d\mathbf{Q}_m^{(n(\nu),k(\nu))} \right| \to 0$$

as $\nu \nearrow \infty$. In fact, the first member of the right-hand side of (6.48) becomes arbitrarily small for M large either because of the supposed uniform integrability of $\log(g_\nu^{(m)})$, $\nu \in I\!N$, w.r.t. Q^m, or because of Lemma 6.3.6 as a consequence of assumption (6.21). The fourth member of the right-hand side of (6.48) can be made small for large M because of (6.47) following from Proposition 6.3.1. In case of the second and the third member of the right-hand side of (6.48) we refer to (6.27) and (6.32), respectively.

Finally we obtain (6.17) through (6.46) with $g_\nu^{(m)}$ in (6.34) by means of

$$| H(Q^m \mid \mathbf{Q}_m^{(n(\nu),k(\nu))}) |$$

$$\leq m \mid H(Q \mid Q_{\varepsilon,k(\nu)}) \mid + \mid \int \log(g_\nu^{(m)}) \, dQ^m - \int \log(g_\nu^{(m)}) \, d\mathbf{Q}_m^{(n(\nu),k(\nu))} \mid$$

$$+ \mid H(\mathbf{Q}_m^{(n(\nu),k(\nu))} \mid Q_{\varepsilon,k(\nu)}^m) \mid \to 0$$

as $\nu \nearrow \infty$. In fact, the first member of the right-hand side vanishes as $k(\nu)$ tends to infinity because of (5.48), the second member vanishes as ν tends to infinity because of (6.48) and the third member vanishes as shown in (6.31). $\Diamond$

Theorem 6.3.1 has some significance in quantum physics. Following Section 1.4, Chapter 2 and Section 6.5, Schrödinger equations are related to Schrödinger processes and vice versa, as far as solutions exist and the corresponding Schrödinger processes can be constructed. Hence Theorem 6.3.1 yields

Corollary 6.3.1 *In the situation of Theorem 6.3.1, the distribution of a 'typical particle' under the law $\mathbf{Q}^{(n,k)}$ as n and k tend to infinity is determined by the Schrödinger process Q and hence by the related Schrödinger equation. In other words, a Schrödinger equation is a 'Boltzmann equation' for a system of interacting particles represented by a system of interacting diffusion processes $((X_t^{(1)}, \ldots, X_t^{(n)}), \ a \leq t \leq b, \ \mathbf{Q}^{(n,k)})$ as n and k tend to infinity.*

We have assigned 'Boltzmann' to Schrödinger equations in order to place them in the concept of propagation of chaos. Moreover, a Schrödinger process Q can be considered as Gibbs state of the system $\mathbf{Q}^{(n,k)}$ representing a microscopical ensemble as n and k tend to infinity. In fact, Q determines the rate function of the large deviation principle called approximate Sanov property in Theorem 6.2.1. Hence Q is the 'most probable' diffusion process under the given circumstances and consequently, it is the limiting distribution postulated by the fundamental hypothesis of statistical mechanics. This illustrates the analogy to classical statistical mechanics as discussed e.g. in Lanford (1973).

The microscopical ensemble featuring a Schrödinger process Q is not uniquely determined. Thus, the question for microscopical systems arises which yield a given 'Boltzmann equation', i.e., a Gibbs state in terms of a Schrödinger process Q. Theorem 6.3.1 claims the existence of a certain Markovian system $\mathbf{Q}^{(n,k)}$ which is probably the most natural one among all microscopical systems.

Remark 6.3.3 We have experienced various difficulties caused by the non-symmetrical relative entropy-'distance' in (5.26). However, Theorem 6.3.1 provides a comparably strong result in terms of convergence in relative entropy. In fact, following Definition 6.3.1, (5.20) immediately implies that propagation of chaos in entropy yields propagation of chaos in variation and propagation of chaos in variation yields propagation of chaos in weak convergence.

6.4   Renormalization procedures

Let us investigate the probability measure $\overline{P}$ in (6.6) which is defined as renormalization of the process $P^c_{(s,x)}$ in (6.5) with creation and killing c. The probabilistical meaning of $\overline{P}$ is not obvious at all; it is hidden in the renormalization procedure (6.6). Extending tools developed in Chapter 2 and Chapter 3, we will characterize $\overline{P}$ by means of a multiplicative functional and by means of a conditioning on survival. An alternative renormalization of $P^c_{(s,x)}$ will be considered at the end of this section.

Let ξ be defined in (6.4) where the potential c is assumed to be in $L^1_{loc}(D)$. Then the Markov property of $P_{(s,x)}$ yields

$$\xi(s,x) = P_{(s,x)}[\exp\{\int_s^t c(v, X_v)\, dv\}\, \xi(t, X_t)\, 1_{\{t<T_s\}}] \qquad (6.49)$$

for $(s,x) \in D$ in (6.2). Moreover, ξ is a solution of the integral equation

$$\xi(s,x) = P_{(s,x)}(b < T_s) + \int_s^b du\, P_{(s,x)}[c(u, X_u)\, \xi(u, X_u)\, 1_{\{u<T_s\}}] \qquad (6.50)$$

for $(s,x) \in D$. In fact,

$$\int_s^b du\, P_{(s,x)}[c(u, X_u)\, \xi(u, X_u)\, 1_{\{u<T_s\}}]$$

$$= \int_s^b du\, P_{(s,x)}[c(u, X_u)\, \exp\{\int_u^b c(v, X_v)\, dv\}\, 1_{\{b<T_s\}}]$$

$$= P_{(s,x)}[\int_s^b du\, c(u, X_u) \sum_{i=1}^{\infty} \frac{1}{(n-1)!}\, (\int_u^b c(v, X_v)\, dv)^{n-1}\, 1_{\{b<T_s\}}] =$$

$$= P_{(s,x)}[\sum_{i=1}^{\infty} \frac{1}{n!} (\int_s^b c(v, X_v) \, dv)^n \, 1_{\{b < T_s\}}]$$

by Lemma 3.2.2 where $P_{(s,x)}(b < T_s)$ on the right-hand side of (6.50) is just the required term for $n = 0$. Hence ξ can be represented as a sum of a space-time harmonic function and the difference of two potentials. Following Blumenthal-Getoor (1968)'s theorem 4.8, $\xi(t, X_t)$ is right-continuous in $t \in [a, b)$.

In view of a transformation explaining the renormalization procedure (6.6) in terms of ξ in (6.4), we define in analogy to N_s^t in Definition 2.3.2 the multiplicative functional

$$\overline{N}_s^t = \exp\{\int_s^t c(v, X_v) \, dv\} \frac{\xi(t, X_t)}{\xi(s, X_s)} \, 1_{\{t < T_s\}} \qquad (6.51)$$

for $a \le s < t \le b$. As a consequence of (6.49),

$$P_{(s,x)}[\overline{N}_s^t] = 1, \quad s \le t \le b \qquad (6.52)$$

for $(s, x) \in D$. Following Section 2.3 and Clark (1970), $\overline{N}_s^t$ is a continuous exponential martingale. Hence Corollary 2.3.2 claims that the space-time diffusion process $((t, X_t), \, s \le t \le b, \, P_{(s,x)})$ can be transformed by means of $\overline{N}_s^t$ into a conservative space-time diffusion process $((t, X_t), \, s \le t \le b, \, \overline{P}_{(s,x)})$ given by

$$\overline{P}_{(s,x)}[F] = P_{(s,x)}[\overline{N}_s^b F] \qquad (6.53)$$

for bounded, measurable space-time functions F.

Theorem 6.4.1 *Provided (6.4) and $c \in L_{loc}^1(D)$, the renormalized process $\overline{P}$ in (6.6) can be represented as*

$$\overline{P}[F] = \int_{D_a} d\varsigma(x) \, \overline{P}_{(a,x)}[F] = \int_{D_a} d\varsigma(x) \, P_{(a,x)}[\overline{N}_a^b F] \qquad (6.54)$$

for bounded, measurable space-time functions F in terms of the multiplicative functional $\overline{N}_s^t$ in (6.51) and the initial distribution ς. Thus, the renormalized process $\overline{P}$ has the diffusion matrix $\alpha = \sigma^2$ of $(X_t, \, a \le t \le b, \, P_{(a,x)})$ and the additional drift

$$\overline{b}(t, y) = \sigma^2 \, \nabla \log \xi(t, y) \quad \text{for} \quad (t, y) \in D \qquad (6.55)$$

if ξ in (6.4) is jointly continuous and in the Sobolev space $H_{loc}^{1,2}(D)$ of Definition 2.2.1.

Proof. By an application of Theorem 2.3.1, the space-time diffusion process (6.53) receives the drift $\bar{b}$ in (6.55). ◇

Looking for a perhaps more intuitive probabilistical meaning of the transformed process $\overline{P}_{(s,x)}$, let us consider

$$\sigma_m = \inf\{t \in [s,b] : \int_s^t c^+(v, X_v)\, dv > m\, (t-s)\} \tag{6.56}$$

for $m \in I\!N$ which implies

$$\int_s^t (c^+(v, X_v) - m)\, dv \leq 0 \quad \text{for } s \leq t \leq \sigma_m \wedge b.$$

The measure $P_{(s,x)}^-$ in (6.3) represents a killed process in the sense of Itô-McKean (1965) and Dynkin (1965). Now we kill the process $P_{(s,x)}^-$ further on at time σ_m in (6.56) as well as by the Kac functional

$$\exp\{\int_s^{t \wedge \sigma_m} (c^+(v, X_v) - m)\, dv\}, \quad s \leq t \leq b \tag{6.57}$$

and call it $P_{(s,x)}^{(m)}$. If $\{survive\ until\ b\}$ denotes the set of all paths which survive until final time b, then

$$P_{(s,x)}^{(m)}[F\, 1_{\{survive\ until\ b\}}] = P_{(s,x)}^c[F\, \exp\{-m\, (b-s)\}\, 1_{\{b<\sigma_m\}}] \tag{6.58}$$

for non-negative measurable functions F where $P_{(s,x)}^c$ is the process in (6.5) with creation and killing c. Hence the killed process $P_{(s,x)}^{(m)}$ conditioned on $\{survive\ until\ b\}$ is given as

$$P_{(s,x)}^{(m)}[F \mid \{survive\ until\ b\}] = \frac{P_{(s,x)}^c[F\, 1_{\{b<\sigma_m\}}]}{P_{(s,x)}^c[1_{\{b<\sigma_m\}}]}. \tag{6.59}$$

We notice that the integrability condition (6.4) yields

$$1_{\{b<\sigma_m\}} \nearrow 1 \quad \text{as } m \nearrow \infty, \quad P_{(s,x)}^-\text{-a.s.}$$

which also holds $P_{(s,x)}^c$-a.s. Consequently, (6.59) converges to the transformed process $\overline{P}_{(s,x)}[F] = P_{(s,x)}^c/\xi(s,x)$ in (6.53) as m tends to infinity. Thus we have shown

Theorem 6.4.2 *Provided (6.4), the transformed process $\overline{P}_{(s,x)}$ in (6.53) possesses a probabilistical meaning in terms of the step by step killed diffusion process $P^{(m)}_{(s,x)}$ in (6.58) conditioned on survival. Actually,*

$$\overline{P}_{(s,x)}[F] = \lim_{m \nearrow \infty} P^{(m)}_{(s,x)}[F \mid \{survive\ until\ b\}] \tag{6.60}$$

for bounded measurable functions F.

Remark 6.4.1 Dawson-Gorostiza-Wakolbinger (1990)'s potential functions c have creation parts $c^+(t,x)$ which are bounded by $M < \infty$. Then $\sigma_M \wedge b = b$ for $m = M$ and the limit in (6.60) is superfluous.

Let us consider an alternative renormalization procedure for the process $P^c_{(s,x)}$ in (6.5) with creation and killing c. While the renormalized process $\overline{P}$ in (6.6) is defined under the local integrability condition (6.4) on ξ, we now introduce the global integrability condition

$$K \equiv \int_{D_a} d\varsigma(x)\,\xi(a,x) < \infty \tag{6.61}$$

where $d\varsigma(x) = \varsigma(x)\,dx$ with $\varsigma > 0$ on D_a is any fixed probability distribution. Condition (6.61) implies (6.4) for λ^d-a.a. $x \in D_a$. Moreover, it allows us to define an alternative renormalized process $\tilde{P}$ by

$$\tilde{P}[F] = \frac{1}{K} \int_{D_a} d\varsigma(x)\,P^c_{(a,x)}[F] \tag{6.62}$$

for bounded, measurable space-time functions F where $P^c_{(a,x)}$ is the process in (6.5) with creation and killing c. Substituting $\overline{P}$ by $\tilde{P}$ in Section 5.1, the results in Section 6.2 remain valid, if $\xi(a,x)$ in (6.6), (6.7) and (6.10) is replaced by the constant K. Consequently, we can state that

Theorem 6.4.1 holds under the integrability condition (6.61) with

$$\tilde{P}[F] = \frac{1}{K} \int_{D_a} d\varsigma(x)\,\xi(a,x)\,P_{(a,x)}[\overline{N}_a^b\,F]$$

in place of (6.54)
as well as

Theorem 6.4.2 remains valid under the integrability condition (6.61) with

$$\tilde{P}[F] = \lim_{m \nearrow \infty} P^{(m)}_k[F \mid \{survive\ until\ b\}] \tag{6.63}$$

in place of (6.60) where $P_k^{(m)}$ is defined in terms of the killed process $P_{(s,x)}^{(m)}$ in (6.58) by

$$P_k^{(m)}[F] = \int_{D_a} d\varsigma(x) \, P_{(a,x)}^{(m)}[F]. \qquad (6.64)$$

Although the two renormalized processes $\overline{P}$ and $\tilde{P}$ in (6.6) and (6.62), respectively, are slightly different in expression, they yield the same Schrödinger process Q under the conditions (6.8) and (6.61). However, the renormalization (6.6) is natural even under (6.61). As a consequence, we should not simply divide by a constant in order to renormalize a process. We better transform it as proposed by Theorem 6.4.1.

6.5   Conditions on creation and killing

As we experienced in Section 6.2, the integrability condition (6.4) is essential for the existence of Schrödinger processes (6.11). It involves merely the positive part of the potential function c while the negative part c^- supports through killing the definition of $P_{(s,x)}^-$ in (6.3). We say that a measurable function c on D in (6.2) is in the *Kato class* for the space-time process $P_{(s,x)}^-$ if it satisfies

$$\lim_{r \searrow 0} \sup_{(s,x) \in D} P_{(s,x)}^- \Big[\int_s^{(s+r) \wedge b} c^+(t, X_t) \, dt \Big] = 0. \qquad (6.65)$$

The Kato class of potential functions in Schrödinger equations is investigated by Aizenman-Simon (1982) where we also refer to Sturm (1989) and Stummer (1990).

Remark 6.5.1 Condition (6.65) is an integrability condition actually not on the positive part c^+, not on the absolute value $\mid c \mid$, but on $c = c^+ - c^-$. This kind of integrability condition ensures in Nagasawa (1990) the continuity of solutions of Feynman-Kac integral equations (3.6).

Proposition 6.5.1 *If a measurable function c on D in (6.2) is in the Kato class, then*

$$\sup_{(s,x) \in D} P_{(s,x)}^- \Big[\exp\{ \int_s^b c^+(t, X_t) \, dt \} \Big] < \infty. \qquad (6.66)$$

Consequently, the integrability condition (6.4) is satisfied uniformly in (s, x) on D.

Proof. For small $r > 0$, the Kato class criterion (6.65) implies

$$\sup_{(s,x) \in D} P^-_{(s,x)}[\int_s^{(s+r) \wedge b} c^+(t, X_t) \, dt] < 1$$

from which (6.66) follows by means of Hasiminsky (1959)'s lemma as is shown in Stummer (1990). $\Diamond$

Let us consider Schrödinger equations of the form

$$i \frac{\partial \Psi}{\partial t} + \frac{1}{2} \nabla \alpha(t, x) \nabla \Psi + i \beta(t, x) \nabla \Psi - V(t, x) \Psi = 0 \qquad (6.67)$$

on $\mathbb{R}^d$. $\nabla \alpha(t, x) \nabla$ denotes the symmetric α-Laplacian in (2.2) with a bounded, positive definite, symmetric diffusion matrix $\alpha_{ij}(t, x) = (\sigma_{ij}(t, x))^2$, $\beta(t, x)$ represents a vector potential satisfying the gauge condition $\nabla \beta = 0$ and $V(t, x)$ is a scalar potential as they were considered in Section 2.1. We assume the existence of the diffusion process (X_t, $s \leq t \leq b$, $P_{(s,x)}$, $(s, x) \in [a, b] \times \mathbb{R}^d$) associated with the diffusion equation (6.68) below in case of $c \equiv 0$. Novikov (1973) provides a sufficient condition on $\beta(t, x)$ cited in (1.51).

If $\Psi = \exp\{R + iS\}$ is a sufficiently smooth solution of the Schrödinger equation (6.67), then the real-valued function $\varphi = \exp\{R + S\}$ satisfies the diffusion equation

$$\frac{\partial p}{\partial t} + \frac{1}{2} \nabla \alpha \nabla p + \beta \nabla p + cp = 0 \quad \text{on } \{0 < |\Psi| < \infty\} \qquad (6.68)$$

with

$$c = -V - 2 \frac{\partial S}{\partial t} - (\sigma \nabla S)^2 - 2\beta \nabla S. \qquad (6.69)$$

The so-called *reference potential* c in (6.69) induced by Ψ is treated in Section 2.1, where we also refer to Remark 6.5.2. Following Theorem 2.1.1, the adjoint real-valued function $\hat{\varphi} = \exp\{R - S\}$ satisfies the adjoint equation of (6.68) which is given as

$$-\frac{\partial p}{\partial t} + \frac{1}{2} \nabla \alpha \nabla p - \beta \nabla p + cp = 0 \quad \text{on } \{0 < |\Psi| < \infty\}. \qquad (6.70)$$

In Section 1.4, we noticed that the complex conjugate $\overline{\Psi}$ is a solution of the adjoint equation of (6.67). Hence the wave functions $\{\Psi, \overline{\Psi}\}$ correspond to $\{\varphi, \hat{\varphi}\}$ and vice versa.

In the converse situation, we consider Schrödinger multipliers Φ and $\hat{\Phi}$ in Definition 2.4.1 which satisfy weakly the diffusion equations (6.68) on $D(\Phi)$ and (6.70) on $D(\hat{\Phi})$, respectively. They determine a Schrödinger process Q in Definition 2.4.1 with the reference potential c in (2.22). In case of $\mu_a = \hat{\Phi}\,\Phi(a,.)$, $\mu_b = \hat{\Phi}\,\Phi(b,.)$ and $c = -L\Phi/\Phi$ on $D(\hat{\Phi})\cap D(\Phi)$, the Schrödinger process Q can alternatively be obtained in (6.11). Then $R = \frac{1}{2}\log(\hat{\Phi}\,\Phi)$ and $S = \frac{1}{2}\log(\Phi/\hat{\Phi})$ yield a weak solution $\Psi = \exp\{R + iS\}$ of the Schrödinger equation (6.67) with a potential V defined by (6.69) as it was investigated in Section 2.1.

The negative part of the potential c

$$c^- = V^+ + 2(\frac{\partial S}{\partial t})^+ + (\sigma\,\nabla S)^2 + 2(\beta\,\nabla S)^+$$

provides the killing in the definition of $P^-_{(s,x)}$ in (6.3) and does in principle not cause any difficulties to the integrability condition (6.4). Our main concern has to be the positive part of the potential c

$$c^+ = V^- + 2(\frac{\partial S}{\partial t})^- + 2(\beta\,\nabla S)^-$$

for which the integrability condition (6.4) takes the form

$$P^-_{(s,x)}[\exp\{\int_s^b \{V^- + 2(\frac{\partial S}{\partial t})^- + 2(\beta\,\nabla S)^-\}(t,X_t)\,dt\}] < \infty.$$

As a consequence, Proposition 6.5.1 implies

Proposition 6.5.2 *Let $\Psi = \exp\{R + iS\}$ represent any 'weak' solution of the Schrödinger equation (6.67) such that $\partial S/\partial t$ as well as ∇S are well-defined. If*

$$\lim_{r\searrow 0}\sup_{(s,x)\in D} P^-_{(s,x)}[\int_s^{(s+r)\wedge b} \{V^- + 2(\frac{\partial S}{\partial t})^- + 2(\beta\,\nabla S)^-\}(t,X_t)\,dt] = 0 \quad (6.71)$$

then the integrability condition (6.4) is satisfied by the reference potential c in (6.69) uniformly in (s,x) on D.

If the potential V belongs additionally to the Kato class (6.65), then condition (6.71) reduces to

$$\lim_{r\searrow 0}\sup_{(s,x)\in D} P^-_{(s,x)}[\int_s^{(s+r)\wedge b} \{(\frac{\partial S}{\partial t})^- + (\beta\,\nabla S)^-\}(t,X_t)\,dt] = 0. \quad (6.72)$$

In particular, if $(\partial S/\partial t)^-$ and $(\beta\,\nabla S)^-$ are bounded, then (6.72) is automatically satisfied. Hence we obtain

Proposition 6.5.3 *Let the potential V be in the Kato class (6.65). If there exists a 'weak' solution $\Psi = \exp\{R + i\,S\}$ of the Schrödinger equation (6.67) such that $(\partial S/\partial t)^- + (\beta\,\nabla S)^-$ satisfies (6.72), then the integrability condition (6.4) is satisfied by c in (6.69) uniformly in (s, x) on D.*

For time independent scalar potential functions V in (6.67) we rediscover

Proposition 6.5.4 (Carmona 1985) *Provided $\sigma \equiv 1$, $\beta \equiv 0$ and a time-independent potential V in the Kato class (6.65), there exists a 'weak' solution $\Psi = \exp\{R + i\,S\}$ of the Schrödinger equation (6.67) such that $\partial S/\partial t$ and ∇S are well-defined. Moreover, the integrability condition (6.4) is satisfied by c in (6.69) uniformly in (s, x) on D, if $(\partial S/\partial t)^-$ is in the Kato class (6.65), in particular if it is bounded. As a consequence, Schrödinger processes Q in (6.11) related to this kind of Schrödinger equations exist.*

As an example we discuss the Schrödinger equation (6.67) with $\sigma \equiv 1$, $\beta \equiv 0$ and a time-independent potential V in the Kato class (6.65). Let $\Psi(t, x) = \exp\{R(x) + i\,E\,t\}$ be any stationary solution, i.e., E is an eigenvalue of the associated Schrödinger operator. In this case the reference potential given by $c(x) = -V(x) - 2E$ satisfies the assumptions in Proposition 6.5.4. It has a singularity of the same order as the potential V but with the opposite sign. Hence, if $V(x) \searrow -\infty$ as $\mid x \mid \searrow 0$ such as the Coulomb potential, then $c(x) \nearrow \infty$ as $\mid x \mid \searrow 0$ which induces intensive creation of particles near the origin under the process $P^c_{(s,x)}$ in (6.5) with creation and killing c. Hence we experience forward prediction. On the other hand, Theorem 6.4.1 claims that there is a strong drift toward the origin under the renormalized process $\overline{P}$ in (6.6) which represents backward prediction. This is nothing but another way of looking at the creation of particles which we have found under the measure $P^c_{(s,x)}$. In fact, following Theorem 6.4.2, we are now tracing backward all particles which were created and which survived until the final time b. In contrast, the performance of the Schrödinger process Q looks completely different from the previous two situations corresponding either to $P^c_{(s,x)}$ or $\overline{P}_{(s,x)}$ in (6.53). The reason is the specific conditioning at initial as well as final time which provides the Eddington-Schrödinger prediction. It induces the typical drift coefficient $b(x) = \nabla R(x)$ which depends only on the stationary distribution $d\mu_a(x) = \exp\{2R(x)\}\,dx = d\mu_b(x)$ of the process. We notice that the Schrödinger process Q is *segregated* by the zero set of $\exp\{R(x)\}$ in excited states. In fact, Theorem 2.4.1 claims that a Schrödinger process Q can neither reach nor cross that zero set. This fact causes serious troubles to any kind of single particle-interpretation of Schrödinger equations and leads naturally to the statistical picture of quantum mechanics proposed by Corollary 6.3.1.

Remark 6.5.2 Kac (1951) investigates the Schrödinger equation (6.67) with $\alpha \equiv 1, \beta \equiv 0$ and a time-independent potential function V. Applying Feynman's path integral formula to the d-dimensional Brownian motion with it in place of t, he concludes that the diffusion equation

$$\frac{\partial p}{\partial t} = \frac{1}{2} \triangle p + V(x)\, p \tag{6.73}$$

should be considered as the real-valued counterpart to the Schrödinger equation (6.67). This approach is not successful since the diffusion equation (6.73) does not give any reasonable physical description of the Schrödinger equation (6.67). A correct answer is provided by our investigations which are based on Schrödinger (1931, 1932): The real-valued counterpart to a Schrödinger equation is a diffusion equation (6.68) which possesses the reference potential c given in (6.69) and which satisfies initial and final conditions in terms of $d\mu_a(x) = \overline{\Psi}(a,x)\, \Psi(a,x)\, dx$ and $d\mu_b(x) = \overline{\Psi}(b,x)\, \Psi(b,x)\, dx$, respectively.

Chapter 7

Schrödinger Systems

The solution of Schrödinger's system (1.17) & (1.18) of non-linear integral equations determines the rate function (1.10) of the large deviation principle in (1.9). In Chapter 6, the large deviation principle called approximate Sanov property is associated with the Schrödinger system in Lemma 6.2.1. Its ingredients are a kernel and prescribed marginal distribution densities. The kernel is represented by the reference process $\overline{P}$ which is given in terms of an initial distribution density $\varsigma(x)$ and a transition density $p(s, x; t, y)$, $s < t$, $x, y \in \mathbb{R}^d$, provided by Theorem 2.3.1.

As investigated in Chapter 2, diffusions related to Schrödinger equations possess transition functions with in general singular creation and killing. Thus they yield generally unbounded kernels which may vanish on sets with positive measure and which can have infinite mass.

The aim of this chapter is to show that Schrödinger systems with such general kernels possess unique solutions in terms of product measures. In Section 7.2, Schrödinger systems are uniquely represented in terms of product measure endomorphisms. Their local adjoints are determined in Section 7.3 by means of a variational principle with constructive appeal. Section 7.4 yields unique solutions with in general unbounded factors which are characterized by integrability properties.

7.1   Non-linear integral equations

A huge cloud of independent, identical particles is observed at initial time a as well as final time b, $-\infty < a < b < \infty$. These two possibly exceptional observations are supposed to be represented in terms of probability densities μ_a and μ_b on $I\!R^d$, respectively. Fixing a and b, we assume that the probabilities of individual particle transitions are given by a kernel p on $I\!R^d \times I\!R^d$. The problem is now to find the 'most probable' kernel p_{cloud} with the observations μ_a and μ_b as prescribed marginal distribution densities.

Let us formulate Schrödinger (1931)'s maximum likelihood result discussed in Section 1.3 for the continuous situation of infinitesimal cell size. In the limit of infinitely many particles we find that $p_{cloud} = \hat{\varphi} \, p \, \varphi$ where $(\hat{\varphi}, \varphi)$ is a pair of non-negative functions determined by

$$\hat{\varphi}(x) \int p(x,z)\,dz\,\varphi(z) \;=\; \mu_a(x)$$

$$\int \hat{\varphi}(x)\,dx\,p(x,z)\,\varphi(z) \;=\; \mu_b(z) \tag{7.1}$$

for λ^d-a.a. $x \in \{\mu_a > 0\}$ and $z \in \{\mu_b > 0\}$, respectively. The system (7.1) of non-linear integral equations was introduced as Schrödinger system in Section 1.3. Existence and uniqueness of its solutions $(\hat{\varphi}, \varphi)$ have already been investigated by means of the large deviation principle in Chapter 6. The rate function in Theorem 6.2.1 results from minimizing the relative entropy expression $H(\,.\,\mid p)$ in (5.6) on the set of probability measures on $I\!R^d \times I\!R^d$ with marginal densities μ_a and μ_b. It is given by

$$H(Q \mid \overline{P}) = \int \log(\frac{dQ}{d\overline{P}})\,dQ$$
$$= \int \log(\hat{\varphi}\,\varphi)\,dQ \;=\; \int \log \hat{\varphi}(x)\,dQ(x,\,.\,) + \int \log \varphi(z)\,dQ(\,.\,,z)$$
$$= \int dx \int dz\,\mu_a(x)\,\mu_b(z)\,\log(\hat{\varphi}(x)\,\varphi(z))$$

where Q is Csiszar's projection in (5.18) of the reference measure $\overline{P}$ with associated kernel p.

Schrödinger (1931) deduces the system (7.1) while searching for real-valued counterparts to Schrödinger equations. Nagasawa (1989) finally concludes: A Schrödinger equation is equivalent to a pair of adjoint non-linear diffusion equations. Their drift coefficients $\nabla \log \hat{\varphi}(t, y)$ and $\nabla \log \varphi(t, y)$ are given by $\hat{\varphi}(t, y) = \int \hat{\varphi}(x) \, dx \, p(a, x; t, y)$ and $\varphi(t, y) = \int p(t, y; b, z) \, dz \, \varphi(z)$, respectively, where the pair $(\hat{\varphi}, \varphi)$ is a solution of the Schrödinger system $(p = p(a, \, . \, ; b, \, . \,), \mu_a = \overline{\Psi}_a \Psi_a, \mu_b = \overline{\Psi}_b \Psi_b)$. Typically, the transition density $p(t_1, y_1; t_2, y_2)$ possesses singular creation and killing related to the solution Ψ of the Schrödinger equation (2.5) as shown in Chapter 2. Consequently, we should be able to deal with (7.1) in case of general kernels as they are provided by transition densities with singular creation and killing evaluated at initial time a and final time b.

In this chapter we intend to establish uniquely and constructively solutions of Schrödinger systems (7.1) determined by (p, μ_a, μ_b) for generally unbounded, jointly measurable kernels p which may vanish on sets with positive Lebesgue measure and which can possess infinite mass. Based on Beurling (1960)'s particular approach, we are going to show that (7.1) has a unique non-negative solution Ξ in terms of a product measure with Lebesque density $d\Xi/d\lambda^{2d} = \hat{\varphi} \, \varphi$ and possibly infinite mass. If M is the probability measure with Lebesgue density $\mu_a \, \mu_b$, then the only required assumption will be

$$M[| \log p | \vee p] < \infty$$

i.e., the finiteness of the expectation of $| \log p | \vee p$ with respect to M which is a condition straightforward to check. The integral $M[\log p]$ is just the term describing the interaction between dynamics p and marginal constraints μ_a, μ_b which also appears in the large deviation approach of Chapter 5. In fact, the rate function $H(Q \mid \overline{P})$ requires the existence of Csiszar's projection Q. Csiszar (1975) gives, besides a condition like (5.3) which is difficult to verify because of its generality, a computable criterion: the finiteness of

$$H(M \mid \overline{P}) = \int \log(\frac{\mu_a \, \mu_b}{p}) \, dM$$
$$= \int \log(\mu_a \, \mu_b) \, dM - \int \log p \, dM = H(M \mid \lambda^d) - M[\log p]$$

which is part of our assumption plus the finite entropy of M. While the present approach assumes $M[p] < \infty$, it does not require an integrability condition on the kernel p exclusively. This is in contrast to the relative entropy method where it is essential that p has finite mass.

In Section 7.2 a setting is developed and the results are given. Section 7.3

deals with a non-linear product measure endomorphism, its local adjoint and
a related variational principle. In Section 7.4 the solution of the Schrödinger
system (7.1) is constructed in terms of a product measure Ξ.

Beurling (1960) generalizes the 2-dimensional Schrödinger system (7.1)
to finitely many dimensions. But at the same time, he restricts to systems
with jointly continuous kernel p satisfying $0 < k_1 \leq p \leq k_2 < \infty$ for constants k_i, $i = 1, 2$. Beurling shows that his assumptions are necessary and
sufficient in order to give the solution Ξ of the Schrödinger system (7.1) by
means of an automorphism on the space of product measures with finite mass.
For our purposes, his approach might not look so promising at first sight. In-
deed, $0 < k_1 \leq p$ seriously restricts applications in probability theory, Gaussian
kernels as well as kernels with singular killing do violate this constraint. Moreover, $p \leq k_2 < \infty$ excludes kernels p with singular creation. Nevertheless, we
will gain considerable benefit from Beurling (1960)'s strategy of defining the
local adjoint of a non-linear product measure endomorphism along the lines of
classical functional analysis.

Solutions $(\hat{\varphi}, \varphi)$ of the Schrödinger system (7.1) will be investigated in
terms of product measures Ξ, $d\Xi = \hat{\varphi}\,\varphi\,d\lambda^{2d}$. The main reason is the uniqueness
of solutions of (7.1) which will mean uniqueness as measures on $\mathbb{R}^d \times \mathbb{R}^d$ as
will be discussed in Remark 7.2.1. First we determine $D = \{0 < \mu_a\,\mu_b < \infty\}$ as
the subset of $\mathbb{R}^d \times \mathbb{R}^d$ where well-defined, non-trivial solutions Ξ of (7.1) can
be expected and call it the relevant domain of solutions Ξ of the Schrödinger
system (7.1). The set D serves as the domain of bounded, measurable test
functions γ suitably chosen as $\gamma(x, z) = \gamma_a(x)\,\gamma_b(z)$ or $\gamma(x, z) = \gamma_a(x) + \gamma_b(z)$
for $(x, z) \in D$.

We will show that the Schrödinger system (7.1) can be expressed by means
of a product measure endomorphism S as

$$S(\Xi) = M\,, \quad dM = \mu_a\,\mu_b\,d\lambda^{2d}$$

where

$$S(\Xi)(\gamma) = \int_D (d\Xi/d\lambda^{2d})\,(d\Psi/d\lambda^{2d})\,\gamma\,d\lambda^{2d}$$

for a density $d\Psi/d\lambda^{2d} = \hat{\psi}\,\psi$ which actually depends on $d\Xi/d\lambda^{2d} = \hat{\varphi}\,\varphi$. Following Beurling (1960), we call Ψ, $d\Psi = \hat{\psi}\,\psi\,d\lambda^{2d}$, the local adjoint of S. Under
Beurling (1960)'s assumptions, the density $\hat{\psi}\,\psi$ is continuous and uniformly

bounded on $\mathbb{R}^d \times \mathbb{R}^d$ with a positive lower bound and S turns out to be an automorphism. Under our assumptions which are chosen to be more appropriate to the actual meaning of Schrödinger systems (7.1), $\hat{\psi}\psi$ is merely a positive, generally unbounded density well-defined on D. There exists a generally unbounded measurable function $\theta(x, z) = \theta_a(x) + \theta_b(z)$, $(x, z) \in D$, such that $(\hat{\psi}\psi)^{-1} = \exp\{\theta\}$.

Our approach adopts Beurling (1960)'s strategy for the construction of solutions Ξ of (7.1) in the representation

$$d\Xi = \exp\{\theta\} \, \mu_a \, \mu_b \, d\lambda^{2d}. \tag{7.2}$$

Beurling (1960)'s assumptions on regularity, uniform boundedness and integrability are reduced to measurability on the relevant domain D, $M[p] < \infty$ and Beurling (1960)'s integrability assumption $M[\|\log p\|] < \infty$. As a consequence, our modified assertions require refined proofs. In particular, we have to overcome the lack of non-trivial bounds for the kernel p. Moreover, new problems arise from the possibility of singular solutions Ξ which will be characterized by means of integrability properties. It happens the actual discovery that Beurling (1960)'s lemma 2 is more universal. In fact, it can be established under our weak assumptions rather naturally. As a consequence, the so-called local linear infimum points $\theta^{(n)}$, $n \in \mathbb{N}$, of the variational functional $V(\gamma, M) = M[p \exp\{\gamma\} - \gamma]$ provide a Cauchy sequence $dM \, p \, \exp\{\theta^{(n)}\}$ on the unit sphere in the Banach space of bounded linear functionals. The limit R is identified as a probability measure and $d\Xi = p^{-1} \, dR$ is proved to solve $S(\Xi) = M$. Moreover, there exists a subsequence of $(\theta^{(n)})_{n \in \mathbb{N}}$ which converges to θ, λ^{2d}-a.e. on D, where θ appearing in (7.2) turns out to be the unique global infimum point of V over functions $\gamma = \gamma_a + \gamma_b$ on D. Finally we show that $M[\theta]$ as well as $M[\log(\hat{\varphi}/\mu_a)]$ and $M[\log(\varphi/\mu_b)]$ are finite. A corollary is devoted to the approximation of Ξ by means of $\Xi^{(n)}$, $d\Xi^{(n)} = dM \, \exp\{\theta^{(n)}\}$, $n \in \mathbb{N}$.

7.2 Product measure endomorphisms

Let $(\mathbb{R}^d, \sigma_{Borel}(\mathbb{R}^d), \lambda^d)$ be the state space at initial time a and final time b, respectively. We are about to investigate non-negative product measures Π on $(\mathbb{R}^{2d}, \sigma_{Borel}(\mathbb{R}^{2d}), \lambda^{2d})$ with $d\Pi = \pi \, d\lambda^{2d} = \pi_a \pi_b \, d\lambda^{2d}$. This style of notation will be used consistently; in particular, π_a and π_b will denote functions defined on subsets of the state space $\mathbb{R}^d$ at initial time a and final time b, respectively. Despite technically problematical in case of measures Π with infinite mass, we would like to think of π_a and π_b as marginal densities at initial and final time, respectively, in order to keep the relation to the probabilistical background of Schrödinger systems (7.1). However, it is preferred to say non-negative, generally unbounded kernel instead of initial density times transition density with in general singular creation and killing evaluated simultaneously at initial time a and final time b. In fact, there are not any intermediate times $t \in (a, b)$, so the Chapman-Kolmogorov property usually associated with a transition density does not play any role. We rather intend to consider the given kernel p as the density of a reference measure on $(\mathbb{R}^{2d}, \sigma_{Borel}(\mathbb{R}^{2d}), \lambda^{2d})$. For technical reasons, i.e., in order to show in Section 7.4 that the constructed solution Ξ of (7.1) is a product measure, we have to restrict to kernels p which have the following property:

Definition 7.2.1 A kernel p on $\mathbb{R}^d \times \mathbb{R}^d$ is called locally uniformly positive if for every $(x_0, z_0) \in \{0 < p < \infty\}$ there exist an open ball $B(x_0, z_0)$ centered at (x_0, z_0) with $B(x_0, z_0) \subset \{0 < p < \infty\}$ and a constant $\delta(x_0, z_0) > 0$ such that $p(x, z) > \delta(x_0, z_0), \forall (x, z) \in B(x_0, z_0)$.

We notice that if a kernel p is jointly continuous on $\{0 < p < \infty\}$ then it is locally uniformly positive.

Theorem 7.2.1 *Let p be a locally uniformly positive kernel in Definition 7.2.1 which may vanish on sets with positive λ^{2d}-measure and which can possess infinite mass. Let the product measure M, $dM = \mu_a \mu_b \, d\lambda^{2d}$, be given by probability densities μ_a and μ_b with $\{0 < \mu_a \mu_b < \infty\} \subset \{0 < p < \infty\}$ such that*

$$M[|\log p| \vee p] < \infty. \tag{7.3}$$

Then there exists a measure-theoretically unique *non-negative product measure Ξ, $d\Xi = \hat{\varphi} \varphi \, d\lambda^{2d}$, concentrated on*

$$D = \{0 < \mu_a \mu_b < \infty\}$$

which is equivalent to M, which has in general infinite mass and which factors $\hat{\varphi}$ and φ solve the Schrödinger system (7.1) determined by (p, μ_a, μ_b) for λ^d-a.a. $x \in \{\mu_a > 0\}$ and $z \in \{\mu_b > 0\}$, respectively. These two factors, only uniquely *determined up to multiplicative constants, satisfy*

$$\mid M[\log(\hat{\varphi}/\mu_a)] \mid < \infty \quad and \quad \mid M[\log(\varphi/\mu_b)] \mid < \infty. \tag{7.4}$$

Corollary 7.2.1 *In the situation of Theorem 7.2.1, let* $(\theta^{(n)} = \theta_a^{(n)} + \theta_b^{(n)})_{n \in \mathbb{N}}$ *be any sequence of bounded functions defined on* D *with*

$$M[p \exp\{\theta^{(n)}\}] = 1, \quad \forall n \in \mathbb{N}$$

such that

$$M[\theta^{(n)}]$$

is monotonously increasing as n *tends to infinity. Then there exists the limit*

$$dR = \lim_{n \nearrow \infty} p \exp\{\theta^{(n)}\} \, dM$$

in the norm of bounded linear functionals. It yields the unique solution Ξ *of the Schrödinger system (7.1) as*

$$d\Xi = p^{-1} \, dR \quad on \;\; D.$$

The product measure Ξ *can be represented as*

$$d\Xi = \exp\{\theta_a\} \exp\{\theta_b\} \, dM$$

where $\theta = \theta_a + \theta_b$ *on* D *has finite* $M[\theta]$. *Moreover, there exists a subsequence of* $(\theta^{(n)})_{n \in \mathbb{N}}$ *which converges to* θ, λ^{2d}-*a.e. on* D, *where*

$$M[\theta^{(n)}] \nearrow M[\theta] \quad as \; n \nearrow \infty.$$

For the actual computation

$$\Xi[\gamma] = \lim_{n \nearrow \infty} M[\exp\{\theta^{(n)}\} \gamma]$$

is well-defined for any non-negative, measurable test function γ *on* D.

The proof of Theorem 7.2.1 dealing with the existence of Ξ is given by means of a construction in Section 7.4. The integrability property (7.4) is provided by Lemma 7.3.4. A variational principle investigated in Section 7.3 yields the approximation properties of Ξ formulated in Corollary 7.2.1. They are deduced in Proposition 7.3.1 and Proposition 7.3.2.

Remark 7.2.1 It is immediate from the Schrödinger system (7.1) that $d\Xi\, p$ has to be a probability measure; consequently $\Xi \in L^1(p)$. Moreover, with any solution $(\hat{\varphi}, \varphi)$, $(k^{-1}\hat{\varphi}, k\,\varphi)$, k a constant, is also a solution of the Schrödinger system (7.1). Hence it is sensible to deal with product measures $d\Xi = \hat{\varphi}\,\varphi\,d\lambda^{2d}$ instead of $(\hat{\varphi}, \varphi)$. Investigations under relevant assumptions adopted from the theory of Schrödinger equations in Chapter 2 show that solutions $(\hat{\varphi},\ \varphi)$ of (7.1) will typically lead in the present approach to product measures Ξ which have infinite mass. In case of signed measures Ξ, their integrals would a priori not be well-defined. Hence and also in accordance with the probabilistical background, we restrict in advance our search for solutions Ξ of (7.1) to non-negative product measures. As a consequence, $\hat{\varphi}$ and φ are considered to be non-negative.

The factorization of product measures is ambiguous in general. Hence we consider

Definition 7.2.2 In case of non-negative product measures Π with finite mass, the factorization obeying $\Pi_r[1] = \sqrt{\Pi[1]}$, $r = a, b$, is called *normalized*.

Clearly any two measures Π_a and Π_b define a unique product measure Π by $d\Pi = d\Pi_a\, d\Pi_b$. Conversely, any two factors $k^{-1}(x)\, d\Pi_a(x)$ and $k(z)\, d\Pi_b(z)$ with $k^{-1}(x)\, k(z) = 1$, $\forall (x, z)$, i.e., k a constant, produce the same product measure Π. In case Π is a non-negative product measure with $0 < \Pi[1] < \infty$, we obtain a unique k by using normalized factors, i.e., $k = \Pi_b^{-1}[1]\, \sqrt{\Pi[1]}$.

Lemma 7.2.1 *Let*

$$\hat{\psi}(x) \;=\; \int p(x, z)\, dz\, \varphi(z) \quad for\ \lambda^d\text{-}a.a.\ x \in \mathbb{R}^d$$

$$(7.5)$$

$$\psi(z) \;=\; \int \hat{\varphi}(x)\, dx\, p(x, z) \quad for\ \lambda^d\text{-}a.a.\ z \in \mathbb{R}^d.$$

Then the Schrödinger system (7.1) appears as

$$\begin{aligned}
\hat{\varphi}(x)\, \hat{\psi}(x) &= \mu_a(x) \quad for\ \lambda^d\text{-}a.a.\ x \in \mathbb{R}^d \\
\varphi(z)\, \psi(z) &= \mu_b(z) \quad for\ \lambda^d\text{-}a.a.\ z \in \mathbb{R}^d
\end{aligned} \qquad (7.6)$$

and the relevant domain of its solutions Ξ, $d\Xi = \hat{\varphi}\,\varphi\,d\lambda^{2d}$, is given as

$$D = \{0 < p < \infty\} \cap \{0 < \mu_a\, \mu_b < \infty\}. \qquad (7.7)$$

Proof. (7.6) is quite obvious. In order to determine the relevant domain of Ξ, we set

$$
\begin{aligned}
\{0 < p < \infty\} &= \{(x, z) \in \mathbb{R}^{2d} : 0 < p(x, z) < \infty\} \\
\{0 < p < \infty\}_x &= \{z \in \mathbb{R}^d : 0 < p(x, z) < \infty\} \\
\{0 < p < \infty\}_z &= \{x \in \mathbb{R}^d : 0 < p(x, z) < \infty\}.
\end{aligned}
$$

Because of (7.5) we may assume without loss of generality that

$$
\{\hat{\varphi} > 0\} = \bigcup_{z \in \mathbb{R}^d} \{0 < p < \infty\}_z, \quad \{\varphi > 0\} = \bigcup_{x \in \mathbb{R}^d} \{0 < p < \infty\}_x.
$$

Hence the density $d\Xi/d\lambda^{2d} = \hat{\varphi}\,\varphi$ needs only to be investigated on

$$
\bigcup_{z \in \mathbb{R}^d} \{0 < p < \infty\}_z \cap \bigcup_{x \in \mathbb{R}^d} \{0 < p < \infty\}_x = \{0 < p < \infty\}.
$$

Since the Schrödinger system (7.6) yields

$$
\{\hat{\varphi}\,\varphi > 0\} \cap \{\hat{\psi}\,\psi > 0\} = \{\mu_a \mu_b > 0\} \tag{7.8}
$$

the relevant domain of $\hat{\varphi}\,\varphi$ is contained in $\{0 < p < \infty\} \cap \{\mu_a \mu_b > 0\}$. As a consequence of (7.6), $\hat{\psi}\,\psi = \infty$ happens if and only if $\mu_a \mu_b = \infty$, provided the convention $\infty \cdot 0 = 0$. $\diamondsuit$

Remark 7.2.2 The relevant domain D in (7.7) of solutions Ξ of Schrödinger systems (7.6) is fairly obvious in the light of the probabilistical background described in Section 7.1. $\{0 < p < \infty\}$ contains all a priori possible random particle transitions from x at time a to z at time b. $\{\mu_a \mu_b = 0\}$ describes those transitions that have not been observed, either because no particle left x at time a, $\mu_a(x) = 0$, or because no particle reached z at time b, $\mu_b(z) = 0$. On $\{\mu_a \mu_b = \infty\}$ the particle traffic jam has already collapsed. Since a particle transition which is a priori not possible will hardly be observed, it is legitimate to assume that $\{0 < \mu_a \mu_b < \infty\} \subset \{0 < p < \infty\}$. Hence we may rely on $D = \{0 < \mu_a \mu_b < \infty\}$ and $M(D) = 1$.

In our treatment of Schrödinger systems (7.6) the appropriate choice of test functions is going to be essential.

Definition 7.2.3 For any subset I of $\mathbb{R}$ let
$\mathcal{F}(D, I)$ denote the set of measurable I-valued functions γ on D,

$\mathcal{S}(D, I) \subset \mathcal{F}(D, I)$ the subset of functions of the form $\gamma = \gamma_a + \gamma_b$ and $\mathcal{P}(D, I) \subset \mathcal{F}(D, I)$ the subset of functions of the form $\gamma = \gamma_a \gamma_b$.

As a consequence of Beppo-Levi's theorem, all integrals $\Pi[\gamma]$ of non-negative, measurable test functions γ with respect to non-negative measures Π with possibly infinite mass are well-defined, also for the integral-value infinity.

Lemma 7.2.2 *The mapping* $S(\Pi)[\gamma] =$

$$= \int \gamma_a(x)\, dx\, (\pi_a(x) \int p(x, z)\, dz\, \pi_b(z)\,) \int \gamma_b(z)\, dz\, (\pi_b(z) \int p(x, z)\, dx\, \pi_a(x)\,)$$

for test functions $\gamma \in \mathcal{P}(D, [0, \infty))$ *is an* endomorphism *of non-negative product measures* Π, $d\Pi = \pi_a\, \pi_b\, d\lambda^{2d}$, *with domain* D.

Proof. The mapping S preserves the product structures as an immediate consequence of its definition. $\Diamond$

Lemma 7.2.3 *Solutions* Ξ, $d\Xi = \hat{\varphi}\, \varphi\, d\lambda^{2d}$, *of the Schrödinger system (7.6) are solutions of*

$$S(\Xi) = M \qquad\qquad (7.9)$$

evaluated on $\mathcal{P}(D, [0, \infty))$, *and vice versa, provided that non-negative product measures with finite mass are considered in their normalized factorization and that their* λ^d-*a.e. equal Lebesgue density factors are identified.*

Proof. Following Lemma 7.2.1, a solution Ξ of (7.6) obviously satisfies (7.9). In case Ξ is a solution of (7.9), $S(\Xi)$ in Lemma 7.2.2 is a non-negative product measure with finite mass which can be factorized as

$$S(\Xi)[\gamma] = k^{-1}\, \Xi[p\, \gamma_a] \cdot k\, \Xi[p\, \gamma_b]\,, \quad \forall\, \gamma \in \mathcal{P}(D, [0, \infty))$$

for arbitrary constants $k > 0$. Because of

$$S(\Xi)[1] = M[1_D] = 1$$

the factors satisfy

$$k^{-1}\, \Xi[p] = S_a(\Xi)[1] = 1\,, \qquad k\, \Xi[p] = S_b(\Xi)[1] = 1$$

from which $k = 1$ follows. Employing normalized factors in Definition 7.2.2, we can identify $dS_a(\Xi)$ and μ_a as well as $dS_b(\Xi)$ and μ_b. $\Diamond$

Remark 7.2.3 Since

$$S(\Xi)[\gamma] = \int dx \int dz \, \gamma(x,z) \, \hat{\varphi}(x) \, \varphi(z) \, \hat{\psi}(x) \, \psi(z) \tag{7.10}$$

is an integral of $\gamma \in \mathcal{P}(D, [0, \infty))$, its definition should be extended linearly as $S(\Xi)[\gamma^{(1)} + \gamma^{(2)}] = S(\Xi)[\gamma^{(1)}] + S(\Xi)[\gamma^{(2)}]$. Consequently, $S(\Xi)$ is given on $\mathcal{S}(D, [0, \infty))$ by means of $\gamma_a + \gamma_b = \gamma_a \cdot 1 + 1 \cdot \gamma_b$ as

$$S(\Xi)[\gamma_a + \gamma_b] = S(\Xi)[\gamma_a] + S(\Xi)[\gamma_b].$$

7.3 A variational principle for local adjoints

In the representation (7.10) the endomorphism S looks like a linear operator determined by the density $\hat{\psi}\psi$ acting on non-negative product measures with density $\hat{\varphi}\varphi$. However, $\hat{\psi}\psi$ depends on $\hat{\varphi}\varphi$ according to (7.5). Hence

Definition 7.3.1 The non-negative product measure Ψ which is determined by $d\Psi = \hat{\psi}\psi \, d\lambda^{2d}$ in (7.5) is called the *local adjoint* of S in Lemma 7.2.2.

Following Lemma 7.2.3, the Schrödinger system (7.6) can be expressed in terms of the local adjoint Ψ of S as

$$\Psi[\varphi \gamma] = M[\gamma], \quad \forall \gamma \in \mathcal{P}(D, [0, \infty)). \tag{7.11}$$

Thus, its solution Ξ is given by

$$\Xi[\gamma] = M[(\hat{\psi}\psi)^{-1} \gamma], \quad \forall \gamma \in \mathcal{P}(D, [0, \infty)) \tag{7.12}$$

where the function

$$(\hat{\psi}\psi)^{-1}(x, z) = \frac{\hat{\varphi}(x)\,\varphi(z)}{\mu_a(x)\,\mu_b(z)} \quad \text{for } (x, z) \in D$$

is in $\mathcal{P}(D, (0, \infty))$. In fact, $(\hat{\psi}\psi)^{-1}$ considered as a multiplication operator is one-to-one and onto on $\mathcal{P}(D, [0, \infty))$ because of (7.8). Hence Ξ and M are equivalent measures on D. Moreover, there exists a unique $\theta \in \mathcal{S}(D, \mathbb{R})$ such that

$$(\hat{\psi}\psi)^{-1}(x, z) = \exp\{\theta_a(x) + \theta_b(z)\} \quad \text{for } (x, z) \in D. \tag{7.13}$$

However, for every constant $k > 0$, there is a decomposition

$$\theta(x, z) = \log(\hat{\varphi}(x)\,(k\,\mu_a(x))^{-1}) + \log(\varphi(z)\,k\,\mu_b^{-1}(z)) \in \mathcal{S}(D, \mathbb{R}).$$

The strategy is now to express Ξ in (7.9) in terms of Ψ in Definition 7.3.1 by relation (7.12). In order to determine Ψ, we are going to deduce a *variational principle* for θ in (7.13).

Lemma 7.3.1 *If*

$$\Xi[\gamma] = M[\exp\{\theta\}\,\gamma]\,, \quad \forall \gamma \in \mathcal{P}(D, [0, \infty))$$

is a solution of the Schrödinger system (7.9), then

$$M[p\,\exp\{\theta\}\,\gamma - \gamma] = 0 \tag{7.14}$$

for all $\gamma \in \mathcal{S}_{bdd}(D, \mathbb{R})$.

Proof. Following Lemma 7.2.3, (7.9) is equivalent to the Schrödinger system (7.6). It is given in terms of the endomorphism S in Lemma 7.2.2 as

$$\Xi[p\,\gamma] = S(\Xi)[\gamma] = M[\gamma]\,, \quad \forall \gamma \in \bigcup_{k>0} \mathcal{S}(D, [-k, \infty)). \tag{7.15}$$

On the other hand, the local adjoint Ψ of S in Definition 7.3.1 in terms of θ in (7.13) provides

$$\Xi[\gamma] = M[\exp\{\theta\}\,\gamma]\,, \quad \forall \gamma \in \mathcal{S}(D, [0, \infty))$$

because of (7.12). As a consequence,

$$\Xi[p\,\gamma] = M[p\,\exp\{\theta\}\,\gamma]\,, \quad \forall \gamma \in \bigcup_{k>0} \mathcal{S}(D, [-k, \infty)) \tag{7.16}$$

although p is generally not in $\mathcal{S}(D, [0, \infty))$. However, (7.6) provides $\Xi \in L^1(p)$. Thus (7.16) which is linear in p follows by means of Stone-Weierstrass' theorem. We notice that (7.16) takes finite values for bounded test functions γ because of $\Xi \in L^1(p)$, while the alternative way based on (7.15) yields

$$\Xi[\gamma] = M[p^{-1}\,\gamma]\,, \quad \forall \gamma \in \mathcal{S}(D, [0, \infty))$$

which can be infinite even for a bounded γ. Thus it is the vanishing difference (7.14) of (7.15) and (7.16) which is well-defined for all $\gamma \in \mathcal{S}_{bdd}(D, \mathbb{R})$.$\diamondsuit$

As we will see, (7.14) is the basic motive for

Lemma 7.3.2 *Provided (7.3), $V(\gamma, M) = M[p \exp\{\gamma\} - \gamma]$ is a non-linear functional well-defined for $\gamma \in \mathcal{F}(D, \mathbb{R})$. It takes a finite value for $\gamma = 0$ and it attains its finite global infimum*

$$v(M) = \inf_{\gamma \in \mathcal{S}(D, \mathbb{R})} V(\gamma, M) > -\infty \tag{7.17}$$

on $\mathcal{S}(D, \mathbb{R})$ at $\gamma = \theta$ given in (7.13). Hence $V(\gamma, M)$ provides a link between the known p and M and the wanted θ which determines Ξ through (7.12).

Remark 7.3.1 There might be the question whether the integrability assumption $M[p] < \infty$ contained in (7.3) is really necessary. First of all we notice that in the large deviation approach, Csiszar (1975)'s restriction to probability measures can be considered as a corresponding assumption. Employing $V(\,.\,, M)$ in Lemma 7.3.2 as a variational functional, we have to guarantee at least one test function $\gamma \in \mathcal{S}(D, \mathbb{R})$ for which $V(\gamma, M)$ is finite. $M[\|\log p\|] < \infty$ implies a finite $V(-\log p, M)$ but $-\log p \notin \mathcal{S}(D, \mathbb{R})$ in general. We could try an approximation procedure by means of Stone-Weierstrass' theorem. However, the non-linearity of $V(\,.\,, M)$ would cause serious difficulties. Moreover, we will notice in the proof of Lemma 7.3.2 that $-\log p$ is just the global infimum point of $V(\,.\,, M)$ on $\mathcal{F}(D, \mathbb{R})$. Hence, approaching $-\log p$ in $\mathcal{S}(D, \mathbb{R})$ in order to find a test function γ with finite V-value actually means solving the variational problem in Lemma 7.3.2 associated with the functional $V(\,.\,, M)$. Finally we mention that $M[p] < \infty$ will just be the appropriate assumption to deal with Definition 7.3.2.

Proof. $V(0, M) = M[p] < \infty$ as well as $V(\gamma, M) > -\infty$, $\forall \gamma \in \mathcal{F}(D, \mathbb{R})$, follow both from (7.3). In fact, $p \exp\{\gamma\} - \gamma = \exp\{\log p + \gamma\} - \gamma \geq 1 + \log p$. The function θ in (7.14) satisfies $M[p \exp\{\theta\}] = 1$. Hence $V(\theta, M) > -\infty$ implies $-\infty \leq M[\theta] < \infty$. In order to prove the finite expectation of θ with respect to M, we consider $\theta \vee k$ for constants k tending to $-\infty$. In fact,

$$M[p \exp\{\theta + \gamma\} - (\theta \vee k + \gamma)] - M[p \exp\{\theta + \gamma\} - (\theta \vee k)]$$
$$= M[p \exp\{\theta\} (\exp\{\gamma\} - 1) - \gamma] \geq M[p \exp\{\theta\} \gamma - \gamma] = 0$$

for all $\gamma \in \mathcal{S}_{bdd}(D, \mathbb{R})$. Because of (7.14), Beppo-Levi's theorem yields

$$V(\theta + \gamma, M) - V(\theta, M) = \tag{7.18}$$

$$
\begin{aligned}
&= \lim_{k \searrow -\infty} \left(M[p \exp\{\theta + \gamma\} - (\theta \vee k + \gamma)] - M[p \exp\{\theta\} - (\theta \vee k)]\right) \\
&= M[p \exp\{\theta\} (\exp\{\gamma\} - 1) - \gamma] \geq 0
\end{aligned}
$$

for all $\gamma \in \mathcal{S}_{bdd}(D, I\!\!R)$. Let κ be any element of the vector space $\mathcal{S}(D, I\!\!R)$ with finite $V(\kappa, M)$. First there exists $\gamma \in \mathcal{S}(D, I\!\!R)$ such that $\kappa = \theta + \gamma$. Second there is a sequence $(\gamma_n)_{n \in I\!\!N} \subset \mathcal{S}_{bdd}(D, I\!\!R)$ such that $V(\theta + \gamma_n, M) \to V(\theta + \gamma, M)$ as $n \nearrow \infty$. From (7.18) we conclude that

$$
V(\theta + \gamma_n, M) \geq V(\theta, M), \quad \forall n \in I\!\!N
$$

which yields

$$
V(\kappa, M) \geq V(\theta, M)
$$

as n tends to infinity. $\Diamond$

In order to approach the global infimum point of the functional $V(\,.\,, M)$, we provide the integrability assumption (7.3), fix a sequence $(\varepsilon_n)_{n \in I\!\!N}$ of positive numbers tending monotonously to zero and consider

Definition 7.3.2 $\theta^{(n)} \in \mathcal{S}_{bdd}(D, I\!\!R), n \in I\!\!N$, is called the n-th *local linear infimum point* of the variational functional $V(\,.\,, M)$ in Lemma 7.3.2 if

$$
\begin{aligned}
V(\theta^{(n)}, M) &\leq V(\theta^{(n)} + \delta, M), \quad \forall \delta \text{ constant} & (7.19) \\
V(\theta^{(n)}, M) &< v(M) + \varepsilon_n. & (7.20)
\end{aligned}
$$

Obviously the global infimum point of $V(\,.\,, M)$ satisfies for all $n \in I\!\!N$ the conditions (7.19) and (7.20) posed on local linear infimum points of $V(\,.\,, M)$. In view of the wanted probability measure $dM\, p \exp\{\theta\}$ in (7.16), we conclude from (7.19) under (7.3) that

$$
0 = \frac{d}{d\delta} V(\theta^{(n)} + \delta, M)\, |_{\delta=0} = M[p \exp\{\theta^{(n)}\}] - 1 \tag{7.21}
$$

$$
0 < \frac{d^2}{d\delta^2} V(\theta^{(n)} + \delta, M)\, |_{\delta=0} = M[p \exp\{\theta^{(n)}\}] \tag{7.22}
$$

where δ is constant. Hence $(dM\, p \exp\{\theta^{(n)}\})_{n \in I\!\!N}$ is a sequence on the unit sphere of bounded linear functionals.

Lemma 7.3.3 (generalization of Beurling (1960)'s lemma 2) *Provided   condition (7.3), any two local linear infimum points* $\theta^{(n)}$, $\theta^{(m)} \in S_{bdd}(D, \mathbb{R})$ *of* $V(.,M)$, $m > n$, *satisfy*

$$M[p \mid \exp\{\theta^{(n)}\} - \exp\{\theta^{(m)}\} \mid] < 2\sqrt{2\,\varepsilon_n} \qquad (7.23)$$

$$\mid M[p \exp\{\theta^{(n)}\}\gamma - \gamma] \mid < 2\sqrt{\varepsilon_n} \qquad (7.24)$$

for all $\gamma \in S(D, \mathbb{R})$ *with* $\| \gamma \|_\infty = \sup_{(x,z)\in D} \mid \gamma(x, z) \mid \leq 1$.

Proof. Dealing with (7.23), an application of Schwarz's inequality to $p\,dM$ yields

$$\{M[p \mid \exp\{\theta^{(n)}\} - \exp\{\theta^{(l)}\} \mid]\}^2 \qquad (7.25)$$
$$\leq \;\; M[p \mid \exp\{\tfrac{1}{2}\theta^{(n)}\} - \exp\{\tfrac{1}{2}\theta^{(l)}\} \mid^2] \, M[p \mid \exp\{\tfrac{1}{2}\theta^{(n)}\} + \exp\{\tfrac{1}{2}\theta^{(l)}\} \mid^2].$$

Since $S_{bdd}(D, \mathbb{R})$ is a linear space, (7.20) provides

$$\varepsilon_n \;>\; \frac{1}{2} V(\theta^{(n)}, M) + \frac{1}{2} V(\theta^{(m)}, M) - V(\frac{1}{2}\,\theta^{(n)} + \frac{1}{2}\,\theta^{(m)}, M)$$
$$= \;\; \frac{1}{2} M[p\,(\exp\{\theta^{(n)}\} + \exp\{\theta^{(m)}\} - 2\,\exp\{\tfrac{1}{2}(\theta^{(n)} + \theta^{(m)})\})]$$
$$= \;\; \frac{1}{2} M[p \mid \exp\{\tfrac{1}{2}\,\theta^{(n)}\} - \exp\{\tfrac{1}{2}\,\theta^{(m)}\} \mid^2].$$

Hence the first factor on the right-hand side of (7.25) is smaller than $2\varepsilon_n$. The second factor on the the the right-hand side of (7.25) can be estimated as

$$M[p \mid \exp\{\tfrac{1}{2}\theta^{(n)}\} + \exp\{\tfrac{1}{2}\theta^{(l)}\} \mid^2]$$
$$= \;\; M[p \exp\{\theta^{(n)}\}] + M[p \exp\{\theta^{(l)}\}] + 2\,M[p \exp\{\tfrac{1}{2}\,\theta^{(n)}\}\,\exp\{\tfrac{1}{2}\theta^{(l)}\}] \;\leq$$
$$\leq \;\; 2 + 2\,\sqrt{M[p \exp\{\theta^{(n)}\}]}\,\sqrt{M[p \exp\{\theta^{(l)}\}]} = 4$$

because of (7.21) satisfied by $\theta^{(n)}$ and $\theta^{(m)}$.

The verification of (7.24) starts from (7.20) which implies

$$\varepsilon_n \;>\; V(\theta^{(n)}, M) - V(\theta^{(n)} + \gamma, M)$$
$$= \;\; M[p \exp\{\theta^{(n)}\}\,(1 - \exp\{\gamma\}) + \gamma]$$

for all $\gamma \in S_{bdd}(D, \mathbb{R})$. In case of $\| \gamma \|_\infty \leq 1$, there exists $\tilde{\gamma} \in S_{bdd}(D, \mathbb{R})$ such that $\gamma = \exp\{\gamma\} - 1 - \tilde{\gamma}$ where $0 \leq \tilde{\gamma} \leq \gamma^2$. Hence (7.21) yields

$$-M[p \exp\{\theta^{(n)}\}\gamma - \gamma] \;\leq$$

$$\leq \quad M[p \exp\{\theta^{(n)}\} (1 - \exp\{\gamma\}) + \gamma] + M[p \exp\{\theta^{(n)}\} \tilde{\gamma}]$$

$$< \quad \varepsilon_n + \| \tilde{\gamma} \|_\infty$$

which is also true for $-\gamma$. Consequently,

$$| M[p \exp\{\theta^{(n)}\} \gamma - \gamma] | < \varepsilon_n + \| \gamma \|_\infty^2$$

where γ has to be replaced by $\sqrt{\varepsilon_n}\, \gamma$ in order to complete the proof. $\Diamond$

A first application of Lemma 7.3.3 yields a relation between the local adjoint Ψ of Definition 7.3.1 and the sequence $(\theta^{(n)})_{n \in I\!N}$ of local linear infimum points in Definition 7.3.2.

Proposition 7.3.1 *First, solutions Ξ of Schrödinger systems (7.9) are unique. Second, if condition (7.3) and a solution Ξ of (7.9) in the representation $d\Xi = dM\, p \exp\{\theta\}$ in (7.12) are provided, then there exists a subsequence of local linear infimum points $\theta^{(n)}$ in Definition 7.3.2 which converges (pointwise) to θ in (7.13), λ^{2d}-a.e. on D.*

Proof. In order to prove the uniqueness of solutions of (7.9), we assume that there are two of them, denoted by $\Xi^{(1)}$ and $\Xi^{(2)}$. Hence there exist $\theta^{(1)}$ and $\theta^{(2)}$ in (7.13) which determine the local adjoints $\Psi^{(1)}$ and $\Psi^{(2)}$ in Definition 7.3.1, respectively. Lemma 7.3.1 provides

$$M[p \exp\{\theta^{(\iota)}\} \gamma - \gamma] = 0, \quad \forall \gamma \in \mathcal{S}_{bdd}(D, I\!R)$$

for $\iota = 1, 2$ which possess the difference

$$M[p\, (\exp\{\theta^{(1)}\} - \exp\{\theta^{(2)}\})\, \gamma] = 0, \quad \forall \gamma \in \mathcal{S}_{bdd}(D, I\!R).$$

Thus the test function

$$
\begin{aligned}
\gamma \;&=\; \frac{\exp\{\theta_a^{(1)}\}}{\exp\{\theta_a^{(1)}\} + \exp\{\theta_a^{(2)}\}} - \frac{\exp\{\theta_b^{(2)}\}}{\exp\{\theta_b^{(1)}\} + \exp\{\theta_b^{(2)}\}} \\[2mm]
&=\; \frac{1}{1 + \dfrac{\exp\{\theta_a^{(2)}\}}{\exp\{\theta_a^{(1)}\}}} - \frac{1}{1 + \dfrac{\exp\{\theta_b^{(1)}\}}{\exp\{\theta_b^{(2)}\}}} \\[2mm]
&=\; \frac{\exp\{\theta_a^{(1)} + \theta_b^{(1)}\} - \exp\{\theta_a^{(2)} + \theta_b^{(2)}\}}{(\exp\{\theta_a^{(1)}\} + \exp\{\theta_a^{(2)}\})\,(\exp\{\theta_b^{(1)}\} + \exp\{\theta_b^{(2)}\})}
\end{aligned}
$$

which evidently is in $\mathcal{S}_{bdd}(D, \mathbb{R})$, yields

$$M[p \, \frac{(\exp\{\theta^{(1)}\} - \exp\{\theta^{(2)}\})^2}{(\exp\{\theta_a^{(1)}\} + \exp\{\theta_a^{(2)}\}) \, (\exp\{\theta_b^{(1)}\} + \exp\{\theta_b^{(2)}\})}] = 0.$$

Now we employ the monotonously increasing subsets

$$B_\delta^{(p)} = \{\delta < p < 1/\delta\} \nearrow \quad \text{as} \quad \delta \searrow 0$$

and

$$B_\rho^{(\iota)} = \{|\, \theta_a^{(\iota)} \,| \le \rho\} \cap \{|\, \theta_b^{(\iota)} \,| \le \rho\} \nearrow \quad \text{as} \quad \rho \nearrow \infty$$

for $\iota = 1, 2$ which satisfy

$$D \subset \{0 < p < \infty\} = \bigcup_{\delta > 0} B_\delta^{(p)}$$

and

$$D = \{0 < \hat{\varphi} \varphi < \infty\} = \bigcup_{\rho > 0} (B_\rho^{(1)} \cap B_\rho^{(2)})$$

because of (7.8). In fact, $(\exp\{\theta_a^{(1)}\} + \exp\{\theta_a^{(2)}\}) \, (\exp\{\theta_b^{(1)}\} + \exp\{\theta_b^{(2)}\})$ is uniformly bounded on $B_\rho^{(1)} \cap B_\rho^{(2)}$, hence

$$M[1_{B_\delta^{(p)}} \, 1_{B_\rho^{(1)} \cap B_\rho^{(2)}} \, (\exp\{\theta^{(1)}\} - \exp\{\theta^{(2)}\})^2] = 0$$

for $0 < \delta, \rho < \infty$. Thus Beppo-Levi's theorem yields $\theta^{(1)} = \theta^{(2)}$, λ^{2d}-a.e. on D.

In order to show the convergence of a subsequence of local linear infimum points $\theta^{(n)}$ to θ in (7.13), we consider the difference of (7.24) and (7.14) which is given as

$$|\, M[p \, (\exp\{\theta^{(n)}\} - \exp\{\theta\}) \, \gamma] \,| < 2\sqrt{\varepsilon_n}, \quad \forall \gamma \in \mathcal{S}(D, \mathbb{R}) \ \text{with} \ \|\, \gamma \,\|_\infty \le 1.$$

As a consequence, the admissible test function

$$\gamma \ = \ \frac{\exp\{\theta_a^{(n)}\}}{\exp\{\theta_a^{(n)}\} + \exp\{\theta_a\}} - \frac{\exp\{\theta_b\}}{\exp\{\theta_b^{(n)}\} + \exp\{\theta_b\}}$$

yields

$$M[p \, \frac{(\exp\{\theta^{(n)}\} - \exp\{\theta\})^2}{(\exp\{\theta_a^{(n)}\} + \exp\{\theta_a\}) \, (\exp\{\theta_b^{(n)}\} + \exp\{\theta_b\})}] \to 0 \quad \text{as} \ n \nearrow \infty.$$

Let

$$B_\rho^{(\theta)} = \{|\,\theta_a\,| \le \rho\} \cap \{|\,\theta_b\,| \le \rho\} \;\nearrow\quad \text{as } \rho \nearrow \infty$$

and let

$$B_\rho^{(n)} = \{|\,\theta_a^{(n)}\,| \le \rho\} \cap \{|\,\theta_b^{(n)}\,| \le \rho\} \;\nearrow\quad \text{as } \rho \nearrow \infty \text{ for } n \in I\!N$$

with

$$B_\rho^{(\cup)} = \bigcup_{n \in I\!N} B_\rho^{(n)}.$$

Then (7.8) implies

$$D = \{0 < \hat\varphi\,\varphi < \infty\} = \bigcup_{\rho > 0} B_\rho^{(\theta)}$$

as well as

$$D = \bigcup_{\rho > 0} B_\rho^{(\cup)}$$

because of Definition 7.3.2. Since $(\exp\{\theta_a^{(n)}\} + \exp\{\theta_a\})\,(\exp\{\theta_b^{(n)}\} + \exp\{\theta_b\})$, $n \in I\!N$, is uniformly bounded on $B_\rho = B_\rho^{(\theta)} \cap B_\rho^{(\cup)}$,

$$\lim_{n \nearrow \infty} M[1_{B_\delta^{(p)}}\,1_{B_\rho}\,(\exp\{\theta^{(n)}\} - \exp\{\theta\})^2] = 0$$

follows for $0 < \delta, \rho < \infty$ and $n \in I\!N$. The Markov inequality yields

$$M(\{|\exp\{\theta^{(n)}\} - \exp\{\theta\}\,| \ge \sqrt\varepsilon\} \cap B_\delta^{(p)} \cap B_\rho)$$
$$\le 1/\varepsilon\, M[1_{B_\delta^{(p)} \cap B_\rho}\,(\exp\{\theta^{(n)}\} - \exp\{\theta\})^2]$$

where the the right-hand side vanishes as n tends to infinity for each $\varepsilon > 0$. Hence $(\theta^{(n)})_{n \in I\!N}$ converges to θ on $B_\delta^{(p)} \cap B_\rho$ in probability with respect to M. The Borel-Cantelli lemma claims that there exists a subsequence $(\theta^{(n^{(1)})})_{n \in I\!N}$ which tends to θ, M-a.s. on $B_\delta^{(p)} \cap B_\rho$. In case of $B_{\delta_2}^{(p)} \cap B_{\rho_2} \supset B_\delta^{(p)} \cap B_\rho$ we know that $(\theta^{(n^{(1)})})_{n \in I\!N}$ converges to θ on $B_{\delta_2}^{(p)} \cap B_{\rho_2}$ in probability with respect to M. Hence there exists a subsequence $(n^{(2)})_{n \in I\!N} \subset (n^{(1)})_{n \in I\!N}$ along which $\theta^{(n^{(2)})}$ converges to θ, M-a.s. on $B_{\delta_2}^{(p)} \cap B_{\rho_2}$. Consequently, we can apply the so-called diagonal procedure. Let $(\delta_m, \rho_m)_{m \in I\!N}$ with

$$B_{\delta_l}^{(p)} \cap B_{\rho_l} \subset B_{\delta_m}^{(p)} \cap B_{\rho_m} \quad \text{for } l < m$$

such that

$$\bigcup_{m \in I\!N} B_{\delta_m}^{(p)} \cap B_{\rho_m} = D$$

and let $(n^{(m)})_{n\in I\!N} \subset (n^{(l)})_{n\in I\!N}$ for $l < m$ such that $(\theta^{(n^{(m)})})_{n\in I\!N}$ converges to θ, M-a.s. on $B_{\delta_m}^{(p)} \cap B_{\rho_m}$. As a consequence, the diagonal procedure claims that $(\theta^{(m^{(m)})})_{m\in I\!N}$ converges to θ, M-a.s. on $D = \cup_{m\in I\!N} B_{\delta_m}^{(p)} \cap B_{\rho_m}$, where we notice that M is equivalent to λ^{2d} on D. $\diamondsuit$

A second application of Lemma 7.3.3 is given by

Proposition 7.3.2 *Let Ξ, $d\Xi = dM\, p\, \exp\{\theta\}$ in (7.12), be a solution of (7.9). Provided (7.3), the expectation of θ in (7.13) with respect to M, i.e., $M[\theta]$, is finite and approached monotonously from below by $(M[\theta^{(n)}])_{n\in I\!N}$.*

Proof. Lemma 7.3.2 claims that θ in (7.13) satisfies

$$M[p\,\exp\{\theta\} - \theta] = v(M) \qquad (7.26)$$

where the global infimum $v(M)$ of $V(\,.\,,M)$ on $\mathcal{S}(D,I\!R)$ in (7.17) is finite under (7.3). Writing

$$\theta = p\,\exp\{\theta\} - (p\,\exp\{\theta\} - \theta)$$

we conclude from (7.26) that $M[\theta] = v(M) - 1$ because of $M[p\,\exp\{\theta\}] = 1$ provided by (7.14).

The local linear infimum points $\theta^{(n)}$, $n \in I\!N$, in Definition 7.3.2 satisfy

$$M[p\,\exp\{\theta^{(n)}\} - \theta^{(n)}] \searrow v(M) > -\infty \qquad (7.27)$$

as n tends to infinity because of (7.20). The boundedness of $\theta^{(n)}$ allows us to write

$$M[p\,\exp\{\theta^{(n)}\} - \theta^{(n)}] = M[p\,\exp\{\theta^{(n)}\}] - M[\theta^{(n)}]. \qquad (7.28)$$

As a consequence of (7.21), $M[p\,\exp\{\theta^{(n)}\}] = 1$ for all $n \in I\!N$. Hence (7.27) and (7.28) imply that $(M[\theta^{(n)}])_{n\in I\!N}$ increases monotonously to $v(M) - 1$. $\diamondsuit$

The third application of Lemma 7.3.3 is to show property (7.4) claimed in Theorem 7.2.1, namely,

Lemma 7.3.4 *Let Ξ, $d\Xi = \hat{\varphi}\varphi\, d\lambda^{2d}$, be a solution of the Schrödinger system (7.9). If (7.3) is provided, then the factors $\hat{\varphi}$ and φ satisfy*

$$\mid M[\log(\hat{\varphi}/\mu_a)] \mid < \infty \quad and \quad \mid M[\log(\varphi/\mu_b)] \mid < \infty.$$

Proof. From (7.13) we receive

$$\theta_a = \log \frac{\hat{\varphi}}{\mu_a} \quad \text{and} \quad \theta_b = \log \frac{\varphi}{\mu_b}.$$

Hence a finite $M[\theta]$ does not imply (7.4). Knowing already that $\theta = \theta_a + \theta_b$ in (7.13) is the global infimum point of $V(.,M)$ in Lemma 7.3.2 on $S(D, \mathbb{R})$, we introduce analogously to (7.19) and (7.20) in Definition 7.3.2 the n-th time a local linear infimum point $\theta_a^{(a,n)} + \theta_b$ and the n-th time b local linear infimum point $\theta_a + \theta_b^{(b,n)}$ for $n \in \mathbb{N}$. This means that $(\theta_a^{(a,n)})_{n\in\mathbb{N}}$ is a sequence of bounded real-valued functions defined on subsets of the state space $\mathbb{R}^d$ at initial time a such that

$$V(\theta_a^{(a,n)} + \theta_b, M) \leq V(\theta_a^{(a,n)} + \delta_a + \theta_b, M), \quad \forall \delta_a \text{ constant} \quad (7.29)$$

$$V(\theta_a^{(a,n)} + \theta_b, M) < v(M) + \varepsilon_n \quad (7.30)$$

where $(\varepsilon_n)_{n\in\mathbb{N}}$ is a sequence as it was used in Definition 7.3.2. Under (7.3), property (7.29) implies that

$$\begin{aligned}
0 &= \frac{d}{d\delta_a} V(\theta_a^{(a,n)} + \delta_a + \theta_b, M) \mid_{\delta_a=0} & (7.31)\\
&= M[p \exp\{\theta_a^{(a,n)} + \theta_b\}] - 1 \\
0 &< \frac{d^2}{d\delta_a^2} V(\theta_a^{(a,n)} + \delta_a + \theta_b, M) \mid_{\delta_a=0} = M[p \exp\{\theta_a^{(a,n)} + \theta_b\}]
\end{aligned}$$

where δ_a is constant. Lemma 7.3.3 adapted to the situation of time a local linear infimum points $\theta_a^{(a,n)} + \theta_b$, $n \in \mathbb{N}$, yields analogously to (7.23) and (7.24) that

$$M[p \mid \exp\{\theta_a^{(a,n)}\} - \exp\{\theta_a^{(a,m)}\} \mid] < 2\sqrt{2\varepsilon_n}, \quad m > n \quad (7.32)$$

$$\mid M[p \exp\{\theta_a^{(a,n)}\}\gamma - \gamma] \mid < 2\sqrt{\varepsilon_n} \quad (7.33)$$

for all $\gamma \in S(D, \mathbb{R})$ with $\parallel \gamma \parallel_\infty \leq 1$.

We notice that $M[\theta_a \vee 0] = \infty$ and $M[\ \theta_a \vee 0] = \infty$ could happen a priori simultaneously. Hence let $\theta \vee k$ for some constants k where $\lim_{k\nearrow-\infty} \theta \vee k = \theta$, λ^{2d}-a.e. on D, because of (7.8). Consequently,

$$\begin{aligned}
\mid M[\theta_a \vee k] \mid &\leq \mid M[\theta_a^{(a,n)}] \mid + \mid M[\theta_a^{(a,n)}] - M[\theta_a \vee k] \mid & (7.34)\\
&\leq \mid M[\theta_a^{(a,n)}] \mid + \mid M[p \exp\{\theta_a^{(a,n)} + \theta_b \vee k\} - (\theta_a^{(a,n)} + \theta_b \vee k)] \\
&\quad - M[p \exp\{\theta_a \vee k + \theta_b \vee k\} - (\theta_a \vee k + \theta_b \vee k)] \mid \\
&\quad + \mid M[p \exp\{\theta_a^{(a,n)} + \theta_b \vee k\}] - M[p \exp\{\theta_a \vee k + \theta_b \vee k\}] \mid.
\end{aligned}$$

Now we are going to apply Beppo-Levi's theorem twice. First Lemma 7.3.2 provides the finite limit

$$\lim_{k \searrow -\infty} M[p \exp\{\theta \vee k\} - \theta \vee k] = M[p \exp\{\theta\} - \theta].$$

Then we conclude by means of (7.34) that

$$
\begin{aligned}
\mid M[\theta_a] \mid &= \lim_{k \searrow -\infty} \mid M[\theta_a \vee k] \mid \\
&\leq \mid V(\theta_a^{(a,n)} + \theta_b, M) - v(M) \mid \\
&+ \mid M[p \exp\{\theta_a^{(a,n)} + \theta_b\}] - M[p \exp\{\theta_a + \theta_b\}] \mid
\end{aligned}
$$

which yields the first part of (7.4) because of (7.14), (7.30), Lemma 7.3.2 and (7.31).◇

7.4 Construction of solutions

Let us assume that the kernel p and the marginal probability densities μ_a and μ_b satisfy the integrability condition (7.3).

Definition 7.4.1 Let $\Xi^{(n)}$, $n \in \mathbb{N}$, denote the non-negative measure with on the set D concentrated Lebesgue density $\hat{\varphi}^{(n)} \varphi^{(n)} = \mu_a \mu_b \exp\{\theta^{(n)}\}$ where $\theta^{(n)} = \theta_a^{(n)} + \theta_b^{(n)}$ is provided by Definition 7.3.2.

We notice that $d\Xi^{(n)} p$, $n \in \mathbb{N}$, is a probability measure as a consequence of (7.21). Moreover, (7.23) yields

$$\| \Xi^{(n)}[p \,.\,] - \Xi^{(m)}[p \,.\,] \| < 2\sqrt{2\,\varepsilon_n} \quad \text{for } m > n$$

which means that $(d\Xi^{(n)} p)_{n \in \mathbb{N}}$ is a Cauchy sequence on the unit sphere in the Banach space of bounded linear functionals. Hence there exists a probability measure R on D such that

$$\| R - \Xi^{(n)}[p \,.\,] \| \to 0 \quad \text{as } n \nearrow \infty \tag{7.35}$$

which allows us to introduce

Definition 7.4.2 Let Ξ denote the non-negative measure with on the set D concentrated density $d\Xi = p^{-1}\, dR$.

Lemma 7.4.1 *In case of a locally positive kernel p in Definition 7.2.1, there exists a non-negative, Borel-measurable function of the form $\hat{\varphi}(x)\,\varphi(z)$ for $(x,z) \in D$ such that Ξ in Definition 7.4.2 can be represented as*

$$d\Xi = \hat{\varphi}\,\varphi\,d\lambda^{2d} \quad \text{on } D.$$

Proof. The measure $\Xi^{(n)}$ in Definition 7.4.1 has the Lebesgue density $\hat{\varphi}^{(n)}\,\varphi^{(n)}$ since $\theta^{(n)}$ in Definition 7.3.2 is in $S_{bdd}(D, \mathbb{R}^d)$. Because of $\Xi \ll \Xi^{(n)}$, the *Radon-Nikodym theorem* provides the non-negative, Borel-measurable density $d\Xi/d\lambda^{2d}$ on D. In order to verify the product form of $d\Xi/d\lambda^{2d}$, we take $\gamma \in \mathcal{F}(D, \mathbb{R}^d)$ with $\| \gamma \|_\infty \leq 1$ and consider for any $\delta > 0$ the test function

$$\gamma_1(x,z) = \frac{\gamma(x,z)}{p(x,z)}\,1_{\{p \geq \delta\}} \quad \text{for } (x,z) \in D.$$

Because of $| \gamma_1 | \leq 1/\delta$, (7.35) implies

$$\int \int p\,(d\Xi/d\lambda^{2d} - \hat{\varphi}^{(n)}\,\varphi^{(n)})\,\delta\,\gamma_1 \to 0 \quad \text{as } n \nearrow \infty$$

which yields

$$\int \int (d\Xi/d\lambda^{2d} - \hat{\varphi}^{(n)}\,\varphi^{(n)})\,\gamma\,1_{\{p \geq \delta\}} \to 0 \quad \text{as } n \nearrow \infty.$$

Now the local uniform positivity of p provides for arbitrary $(x_0, z_0) \in D$ an open ball $B(x_0, z_0) \subset D$ centered at (x_0, z_0) and a constant $\delta(x_0, z_0) > 0$ such that

$$\int \int (d\Xi/d\lambda^{2d} - \hat{\varphi}^{(n)}\,\varphi^{(n)})\,\gamma \to 0 \quad \text{as } n \nearrow \infty$$

for all $\gamma \in \mathcal{F}(B(x_0, z_0), \mathbb{R})$ with $\| \gamma \|_\infty \leq 1$. Consequently, we receive

$$\frac{d\Xi}{d\lambda^{2d}}(x,z) = \lim_{n \nearrow \infty} \hat{\varphi}^{(n)}(x)\,\varphi^{(n)}(z)$$

for λ^{2d}-a.a. $(x,z) \in B(x_0, z_0)$ and hence for λ^{2d}-a.a. $(x,z) \in D$. $\Diamond$

Remark 7.4.1 As a consequence of (7.35), $d\Xi\,p$ is a probability measure for a class of non-negative, generally unbounded kernels p. Hence Ξ cannot be expected to have finite mass. Once the existence of Ξ has been shown, its integral values can simply be computed as

$$\Xi[\gamma] = \lim_{n \nearrow \infty} \Xi^{(n)}[\gamma] = \lim_{n \nearrow \infty} M[\exp\{\theta^{(n)}\}\,\gamma]$$

for $\gamma \in \mathcal{F}(D, [0, \infty))$.

As final step we have to prove that the factors $\hat{\varphi}$ and φ in Lemma 7.4.1 solve the Schrödinger system (7.1). Following Lemma 7.2.1 and Lemma 7.2.3 it remains to show

Lemma 7.4.2 *The non-negative measure Ξ in Definition 7.4.2 is a solution of (7.9).*

Proof. Lemma 7.4.1 provides $d\Xi = \hat{\varphi}\,\varphi\,d\lambda^{2d}$. Following Lemma 7.2.3, the integral relation (7.9) has to be verified on $\mathcal{P}(D, [0, \infty))$. As a consequence of Remark 7.2.3, (7.9) will also hold on $\mathcal{S}(D, [0, \infty))$. Let us consider

$$| S(\Xi)[\gamma] - M[\gamma] | \leq | S(\Xi)[\gamma] - S(\Xi^{(n)})[\gamma] | + | S(\Xi^{(n)})[\gamma] - M[\gamma] | \quad (7.36)$$

for $n \in I\!\!N$. The first difference on the the right-hand side of (7.36) can be estimated as

$$
\begin{aligned}
&| \Xi[p\,\gamma_a]\,[\Xi[p\,\gamma_b] - \Xi^{(n)}[p\,\gamma_b]] | + | \Xi^{(n)}[p\,\gamma_b]\,[\Xi[p\,\gamma_a] - \Xi^{(n)}[p\,\gamma_a]] | \\
\leq\ & \| \gamma_a \|_\infty\, \Xi[p]\, | \Xi[p\,\gamma_b] - \Xi^{(n)}[p\,\gamma_b] | \\
+\ & \| \gamma_b \|_\infty\, \Xi^{(n)}[p]\, | \Xi[p\,\gamma_a] - \Xi^{(n)}[p\,\gamma_a] | \\
\leq\ & 2\,\| \gamma_a \|_\infty \| \gamma_b \|_\infty \| \Xi[p\,.\,] - \Xi^{(n)}[p\,.\,] \| < 4\sqrt{2\varepsilon_n}\, \| \gamma_a \|_\infty \| \gamma_b \|_\infty
\end{aligned}
$$

by means of the endomorphism S in Lemma 7.2.2, its representation in (7.15) as well as the properties (7.21) and (7.23). In order to handle the second difference on the the right-hand side of (7.36), we recall Definition 7.3.1 of the adjoint Ψ of S, use (7.13) and notice that $M[\gamma] = M[\gamma_a]\,M[\gamma_b]$. Then

$$
\begin{aligned}
&| M[p\,\exp\{\theta^{(n)}\}\,\gamma_a]\,M[p\,\exp\{\theta^{(n)}\}\,\gamma_b] - M[\gamma_a]\,M[\gamma_b] | \\
\leq\ & | M[p\,\exp\{\theta^{(n)}\}\,\gamma_a] |\,| M[p\,\exp\{\theta^{(n)}\}\,\gamma_b - \gamma_b] | \\
+\ & | M[\gamma_b] |\,| M[p\,\exp\{\theta^{(n)}\}\,\gamma_a - \gamma_a] | \\
<\ & 4\sqrt{\varepsilon_n}\, \| \gamma_a \|_\infty \| \gamma_b \|_\infty
\end{aligned}
$$

follows by means of Definition 7.4.1 and (7.24). $\diamondsuit$

References

Aebi, R.: NM-transformed α-Diffusion with Singular Drift
doctoral thesis, University of Zürich, 1989

Aebi, R., Nagasawa, M.: Large Deviations and the Propagation of Chaos
for Schrödinger Processes
Probab. Theory Relat. Fields **94**, 53–68 (1992a)

Aebi, R.: Itô's Formula for Non-smooth Functions
Publ. RIMS of Kyoto Univ. **28**, 595–602 (1992b)

Aebi, R.: Diffusions with Singular Drift Related to Wave Functions
Probab. Theory Relat. Fields **96**, 107–121 (1993)

Aebi, R.: Watching Clouds – Prediction from Past and Future
Habilitationsschrift
Philosophisch-naturwissenschaftliche Fakultät der Universität Bern, 1994

Aebi, R.: Propagation of Chaos – the Inverse Problem
Progress in Probability **36**, 1–25 (1995)

Aebi, R.: A Solution to Schrödinger's Problem of Non-Linear Integral Equations
Z. Angew. Math. Phys. (ZAMP) **46**, 772–792 (1995b)

Aebi, R.: An Alternative Representation of Diffusions based on Time-Reversibility
Kyushu Journal of Mathematics **49**, no. 2, 1 – 15 (1995c)

Aebi, R.: Schrödinger's Time-Reversal of Natural Laws
to appear in The Mathematical Intelligencer, Springer-Verlag, New York 1995d

Aebi, R.: Propagation of Chaos in Entropy
to appear in Kodai Mathematical Journal, Tokyo 1996

Aizenman, M., Simon, B.: Brownian Motion and Harnack Inequality
for Schrödinger Operators
Comm. Pure Appl. Math. **35**, 209–273 (1982)

Albeverio, S., Hoegh-Krohn, R.: A Remark on the Connection between
Stochastic Mechanics and the Heat Equation
J. Math. Phys. **15**, 1745–1748 (1974)

Aronszjan, N.: Theory of Reproducing Kernels
Trans. Amer. Math. Soc. **68**, 337–404 (1950)

Arnold, L.: Stochastische Differentialgleichungen
Oldenburg-Verlag, München 1973

Azema, J.: Théorie Générale des Processus et Retournement du Temps
Ann. Scient. Ec. Norm. Sup. **6**, 459–519 (1973)

Azencott, R.: Grandes Déviations et Applications
Ecole d'Eté de Probabilité de Saint Flour VIII, 1–176
Lecture Notes in Math 774, Springer-Verlag, New York 1980

Bauer, H.: Probability Theory and Elements of Measure Theory
Academic Press, New York 1981

Beckenbach, E.F., Bellman, R.: Inequalities
Springer-Verlag, Berlin 1961

Bernstein, S.: Sur les Liaisons entre les Grandeurs Aleatoires
Verhandlungen des Internat. Math. Kongresses Zürich, **1**, 288–309 (1932)

Beurling, A.: An Automorphism of Product Measures
Ann. Math. **72** no. 1, 189–200 (1960)

Billingsley, P.: Convergence of Probability Measures
John Wiley & Sons Inc., New York 1968

Blanchard, P., Golin, S.: Diffusion with Singular Drift Fields
Commun. Math. Phys. **109**, 421–435 (1987)

Blanchard, P., Golin, S., Zheng, W.: Mathematical and Physical Aspects
of Stochastic Mechanics
Lecture Notes in Physics **281**, Springer-Verlag 1987

Blumenthal, R.M., Getoor, R.K.: Markov Processes and Potential Theory
Academic Press, New York 1968

Boltzmann, L.: Vorlesungen über Gastheorie
J.A. Barth Verlag, Leipzig 1896

Boltzmann, L.: Über die sogenannte H-Kurve
Ges. Abh. III, Nr. 128, Math. Ann. **50**, 325 (1898)

Born, M., Jordan, W.: Zur Quantenmechanik
Z. Phys. **34**, 858–888 (1925)

Born, M., Heisenberg, W., Jordan, W.: Zur Quantenmechanik II
Z. Phys. **35**, 557–615

Brox, T.: A One-dimensional Diffusion Process in a Wiener Medium
Ann. Probab. **14**, 1206–1218 (1986)

Carlen, E. A.: Conservative Diffusions
Commun. Math. Phys. **94**, 293–315 (1984)

Carmona, R.: Processus de Diffusion Gouverneé par la Form de Dirichlet
de l'Opérateur de Schrödinger
Sém. de Probabilité XIII, Lecture Notes Math. **721**, 557–569 (1979)

Carmona, R. A.: Probabilistic Construction of Nelson Processes
Taniguchi Symp. PMMP Katata, 55–81 (1985)

Carmona, R. A., Nualart, D.: Nonlinear Stochastic Integrators
Stochastic Monographs, vol. 6
Gordon and Breach Science Publishers, New York 1990

Chentsov, N. N.: Statistical Decision Rules and Optimal Inference
Translations of Mathematical Monographs, vol. 53
American Mathematical Society, Providence, Rhode Island 1982

Chung, K.L., Walsh, J.B.: To Reverse a Markov Process
Acta Math. **123**, 225–251 (1969)

Chung, K.L., Rao, K.M.: Feynman-Kac formula and Schrödinger Equation
Sem. Stoch. Proc. (ed Cinlar, Chung, Getoor), 1–29, Birkhäuser, Basel 1981

Chung, K.L., Williams, R.J.: Introduction to Stochastic Integration
2nd ed., Birkhäuser, Basel 1988

Clark, J.M.C.: The Representation of Functionals of Brownian Motion by Stochastic Integrals
Ann. Math. Stat. **41**, 1281–1295 (1970)

Csiszar, I.: Information-type Measures of Difference of Probability Distributions and Indirect Observations
Studia Sci. Math. Hungar. **2**, 299–318 (1967)

Csiszar, I.: I-Divergence Geometry of Probability Distributions and Minimization Problems
Ann. Probab. **3**, 146–158 (1975)

Csiszar, I.: Sanov Property, Generalized I-Projection and a Conditional Limit Theorem
Ann. Probab. **12**, 768–793 (1984)

Csiszar, I.: An Extended Maximum Entropy Principle and a Bayesian Justification
Bayesian Statistics 2, 83–98, Elsevier Science Publishers B.V. 1985

Csiszar, I.: Why Least Squares and Maximum Entropy?
An Axiomatic Approach to Inference for Linear Inverse Problems.
Ann. of Stat. **19** no. 4, 2032–2066 (1991)

Constantinescu, C.: Topologische Räume
vdf, ETHZ, Zürich 1979

Dawson, D.A., Gärtner, J.: Long-time Fluctuations of Weakly Interacting Diffusions
Technical Report of the Laboratory for Research in Statistics and Probability
Carleton University, 1984

Dawson, D.A., Gärtner, J.: Large Deviations, Free Energy Functional and Quasi-potential
for a Mean Field Model of Interacting Diffusions
Memoirs of the AMS, vol. 78, no. 398

Dawson, D., Gorostiza, L., Wakolbinger, A.: Schrödinger Processes and Large Deviations
J. Math. Phys. **31** (10), 2385–2388 (1990)

Dellacherie, C., Meyer, P.A.: Probabilities and Potential
North Holland Publ. Co. 1978

Deuschel, J.-D., Stroock, D.W.: Large Deviations
Academic Press, New York 1989

Dirac, P.A.M.: The Principle of Quantum Mechanics
Oxford 1930, New York 1958

Doob, J.L.: Stochastic Processes
John Wiley & Sons Inc., New York 1953

Doob, J.L.: Classical Potential Theory and its Probabilistic Counterpart
Springer-Verlag, New York 1984

Dorling, J.: Schrödinger's Original Interpretation of the Schrödinger Equation: rescue attempt
Schrödinger, centenary celebration of a polymath (ed. Kilmister, C.W.), 16–40
Cambridge University Press, Cambridge 1987

Dürr, D., Goldstein, S., Zanghi, N.: Quantum Equilibrium and Origin of Absolute Uncertainty
J. Stat. Phys. **67**, 843–905 (1992)

Durrett, R.: Brownian Motion and Martingales in Analysis
Wadsworth, Belmont 1984

Dynkin, E. B.: Markov Processes, vol. 1 & 2
Springer-Verlag, Berlin 1965

Einstein, A., Podolsky, B., Rosen, N.: Can Quantum-Mechanical Description of Physical
Reality be Considered Complete?
Phys. Rev. **47**, 777–780 (1935)

Faris, W., Simon, B.: Degenerate and Nondegenerate Ground States
for Schrödinger Operators
Duke Math. J. **42**, 559–567

Feller, W.: An Introduction to Probability Theory and Its Applications
3rd ed., John Wiley & Sons Inc., New York 1968

Fényes, I.: Eine wahrscheinlichkeitstheoretische Begründung und Interpretation
der Quantenmechanik
Zeitschrift der Physik, **132**, 81–106 (1952)

Feynman, R.P.: Space-time Approach to Non-relativistic Quantum Mechanics
Reviews of Modern Physics **22**, 367–387 (1948)

Föllmer, H.: An Entropy Approach to the Time Reversal of Diffusion Processes
4th IFIP workshop on stochastic differential equations (ed. Métivier, M., Pardoux, E.)
Lecture Notes in Control and Information, Springer-Verlag, Berlin 1985

Föllmer, H., Wakolbinger, A.: Time Reversal of Infinite Dimensional Diffusions
Stoch. Proc. Appl. **22**, 59–77 (1986)

Föllmer, H.: Time Reversal on Wiener Space
Bibos-Symposium 'Stochastic Processes in Mathematical Physics'
Springer Lect. Notes Math. **1158**, 119–129 (1986)

Föllmer, H.: Random Fields and Diffusion Processes, École d'été de Saint Flour XV–XVII
(1985–87), Springer Lect. Notes Math. **1362** (1988)

Fortet, R.: Résolution d'un Système d'Équation de M. Schrödinger
J. Math. Pures et Appl. IX, 83–95 (1940)

Friedman, A.: Partial Differential Equations of Parabolic Type
Prentice Hall Inc., New York 1964

Friedman, A.: Stochastic Differential Equations and Applications, vol. 1 & 2
Academic Press, New York 1975/76

Fukushima, M.: Dirichlet Forms and Markov Processes
Kodansha Ldt, Tokyo 1980

Fukushima, M., Takeda, M.: A Transformation of Symmetric Markov Processes and
Donsker-Varadhan Theory
Osaka J. Math. **21**, 311–326 (1984)

Getoor, R.K., Glover, J.: Riesz Decomposition in Markov Process Theory
Trans. Amer. Math. Soc. **285**, 107–132 (1984)

Getoor, R.K., Sharpe, M.J.: Naturality, Standardness and Weak Duality for Markov Processes
Z. Wahrsch. Verw. Gebiete **67**, 1–62 (1984)

Gihman, I.I., Skorohod, A.V.: Introduction to the Theory of Random Processes
W.B. Saunders Co., Philadelphia 1969

Gihman, I.I., Skorohod, A.V.: The Theory of Stochastic Processes, vol. 1 & 2
Springer-Verlag, Berlin 1974/75

Girsanov, I.V.: On Transforming a Certain Class of Stochastic Processes
by Absolutely Continuous Substitution of Measures
Theor. Probab. Appl. **5**, 285–301 (1960)

Glimm, J., Jaffe, A.: Quantum Physics, a Functional Integral Point of View
Springer-Verlag, Berlin 1987

Grad, H.: The Many Faces of Entropy
Comm. Pure and Appl. Math. **14**, 323–354 (1961)

Groeneboom, P., Oosterhoff, J., Ruymgaart, F.H.: Large Deviation Theorems
for Empirical Probability Measures
Ann. Probab. **7**, no 4, 553–586 (1979)

Guerra, F., Pavon, M.: Stochastic Variational Principles and Free Energy for
Dissipative Processes
In 'Analysis and Control of Nonlinear Systems' (ed. Byrnes, C.I., Martin, C.F., Seaks, R.E.)
Elsevier Science Publ. B.V., North Holland 1988

Gutkin, E., Kac, M.: Propagation of Chaos and Burger's Equation
SIAM J. Appl. Math. **43**, no 4, 971–980 (1983)

Harris, T.E.: Diffusion with 'Collisions' between Particles
J. Appl. Probab. **2**, 323–338 (1965)

Hasiminsky, R.Z.: On Positive Solutions of the Equation $Au + Vu = 0$.
Theory Probab. its Appl. **4**, 309–318 (1959)

Heisenberg, W.: Über Quanten-theoretische Umdeutung kinetischer
und mechanischer Beziehungen
Zeitschrift für Physik **33**, 879–893 (1925)

Helms, L.L.: Markov Processes with Creation of Mass I, II
Z. Wahrscheinlichkeitstheorie verw. Geb. **7**, 225–234, **15**, 208–218 (1969)

Hunt, G.A.: Markov Processes and Potentials
Illinois J. Math. **1**, 44–93, 316–369 (1957); **2**, 151–213 (1958)

Ikeda, N., Nagasawa, M., Sato, K.: A Time Reversal of Markov Processes with Killing
Kodai Math. Sem., Rep. **16**, 88–97 (1964)

Ikeda, N., Watanabe, S.: Stochastic Differential Equations and Diffusion Processes
North-Holland, New York 1989

Itô, K.: On a Stochastic Integral Equation
Proceedings of the Japan Academy **22**, 32–35 (1946)

Itô, K.: On a Formula Concerning Stochastic Differentials
Nagoya Math. J. **3**, 55–65 (1951)

Itô, K.: Lectures on Stochastic Processes
Tata Institute, Bombay 1961

Itô, K.: Wiener Integrals and Feynman Integral
Proc. Fourth Berkeley Symp. on Math. Stat. and Probab. **2**, 227–238
Univ. California Press, Berkeley 1961

Itô, K., McKean, H.P.: Diffusion Processes and Their Sample Paths
Springer-Verlag, Berlin, 1965

Itô, K., Nisio, M.: On the Convergence of Sums of Independent Banach Space
Valued Random Variables
Osaka J. Math. **5**, 35–48 (1968)

Itô, S.: Fundamental solution of parabolic differential equations
and boundary value problems
Japanese J. Math. **27**, 55–102 (1957)

Jamison, B.: Reciprocal Processes
Z. Wahrscheinlichkeitstheorie verw. Geb. **30**, 65–86 (1974)

Jamison, B.: A Martin Boundary Interpretation of the Maximum Entropy Argument
Z. Wahrscheinlichkeitstheorie verw. Gebiete **30**, 265–272 (1974)

Jamison, B.: The Markov processes of Schrödinger
Z. Wahrscheinlichkeitstheorie verw. Gebiete **32**, 323–331 (1975)

Jammer, M.: The Philosophy of Quantum Mechanics.
The Interpretations of Quantum Mechanics in Historical Perspective
John Wiley & Sons Inc., New York 1974

Kac, M.: On Some Connections Between Probability Theory
and Differential and Integral Equations
Proc. Second Berkeley Symp. on Math. Stat. Probab., 189–215
Univ. California Press, Berkeley 1951

Karlin, S., Taylor, H.M.: Second Course in Stochastic Processes
Academic Press, New York 1981

Kempermann, J.H.B.: On the Optimum Rate of Transmitting Information
Probability and Information Theory, 126–169
Lecture Notes in Mathematics, Springer-Verlag, Berlin (1967)

Khinchin, A.I.: Mathematical Foundations of Information Theory
Dover Publications, New York 1957

Khas'minskii, R.Z.: On Positive Solutions of the Equation $Au + V u = 0$
Theory Probab. Appl. **4**, 309–318 (1959)

Kingman, J.F.C., Taylor, S.J.: Introduction to Measure and Probability
Cambridge University Press, Cambridge 1977

Kolmogoroff, A.: Analytische Methoden in Wahrscheinlichkeitsrechnung
Math. Ann. **104**, 415–458 (1931)

Kolmogoroff, A.: Zur Theorie der Markoffschen Ketten
Math. Ann. **112**, 155–160 (1936)

Kolmogoroff, A.: Zur Umkehrbarkeit der Statistischen Naturgesetze
Math. Ann. **113**, 766–772 (1937)

Krylov, N.V.: Controlled Diffusion Processes
Applications of Mathematics, vol. 14
Springer-Verlag, New York 1980

Kullback, S.: A Lower Bound for Discrimination Information in terms of Variation
IEEE Trans. Information Theory, **IT-13**, 126–127 (1967)

Kunita, H., Watanabe, T.: Notes on Transformations of Markov Processes connected
with Multiplicative Functionals
Memoirs of the Faculty of Science, Kyushu University, Ser. A **17**, 181–191 (1963)

Kunita, H., Watanabe, T.: On Certain Reversed Processes and
Their Applications to Potential Theory and Boundary Theory
J. Math. Mech. **15**, 398–434 (1966)

Kunita, H.: Stochastic Flows and Stochastic Differential Equations
Cambridge University Press, Cambridge 1990

Kusuoka, S., Tamura, Y.: Gibbs Measures for Mean Field Potentials
J. Fac. Sci. Univ. Tokyo, Sec. IA Math. **31**, 223–245 (1984)

Kuznezov, S.E.: Construction of Markov Processes with Random Birth and Death Times
Theory Probab. Appl. **18**, 596–601 (1973)

Lanford, O.E.: Entropy and Equilibrium States in Classical Statistical Mechanics
In 'Statistical Mechanics and Mathematical Problems' (ed. Lenard A.)
Lecture Notes in Phys. **20**, 1–113, Springer-Verlag, Berlin 1973

Liptser, R. S., Shiryayev, A. N.: Statistics of Random Processes I;
Applications of Mathematics, vol. 5, Springer-Verlag, New York 1977

Mackey, G.W.: Unitary Group Representations in Physics, Probability and Number Theory
Benjamin/Cummings Pub. Co. Inc., London 1978

Maruyama, G.: Markov Processes and Stochastic Equations
Natural Science Report; Ochanomizu University **4**, 40–43 (1953)

Maruyama, G.: On the Transition Probability Functions of the Markov Process
Natural Science Report; Ochanomizu University **5**, 10–20 (1954)

Maruyama, G.: Continuous Markov Processes and Stochastic Equations
Rendiconti del Circolo Matematico di Palermo, Ser. II **4**, 48–90 (1955)

McKean, H.P.: A Class of Markov Processes associated with Non-Linear Parabolic Equations
Proc. Natl. Acad. Sci. **56**, 1907–1911 (1966)

McKean, H.P.: Propagation of Chaos for a Class of Nonlinear Parabolic Equations
Lecture Series in Differential Equations, Catholic Univ., 41–57 (1967)

Meyer, P.A.: Fonctionnelles Multiplicatives et Additives de Markov
Ann. Inst. Fourier **12**, 125–230

Meyer, P.A.: La Propriété de Markov Forte des Fonctionelles Multiplicatives
Theory of Probab. its Appl. **8**, 328–334 (1963)

Meyer, P. A.: Probability and Potentials
Blaisdell Pub. Co, Waltham 1966

Meyer, P.A.: Le Retournement du Temps d'après Chung et Walsh
Springer Lect. Notes in Math. **191**, 213–245 (1971)

Meyer, P.A.: Renaissance, Recollement, Mélanges, Ralentissement de Processus de Markov
Ann. Inst. Fourier, Grenoble **25**, 465–497 (1975)

Meyer, P. A., Zheng, W. A.: Quelques Resultats de Mécanique Stochastique
Seminaire de Probabilités **XVIII**, 223–244 (1982/83)

Meyer, P. A., Zheng, W. A.: Sur la Construction des Certaines Diffusions
Seminaire de Probabilités **XX**, 334–337 (1984/85)

Mitro, J.B.: Dual Markov Processes: Construction of Useful Auxiliary Processes
Z. Wahrsch. verw. Gebiete **47**, 139–156 (1979)

Moore, W.J.: Schrödinger, Life and Thoughts
Cambridge University Press, Cambridge 1989

Nagasawa, M.: The Adjoint Process of a Diffusion Process with Reflecting Barrier
Kodai Math. Sem. Rep. **13**, 235–248 (1961)

Nagasawa, M., Sato, K.: Some Theorems on Time change and Killing of Markov Processes
Kodai Math. Sem. Rep. **15**, 195–219 (1963)

Nagasawa, M.: Time Reversions of Markov Processes
Nagoya Math. J. **24**, 177–204 (1964)

Nagasawa, M.: Markov Processes with Creation and Annihilation
Z. Wahrsch. verw. Gebiete **14**, 49–60 (1969)

Nagasawa, M.: Macroscopic, Intermediate, Microscopic and Mesons
Springer Lect. Notes Phys. **262**, 427–437 (1985)

Nagasawa, M., Tanaka, H.: A Diffusion Process in a Singular Meandrift-Field
Z. Wahrsch. verw. Gebiete **68**, 247–269 (1985)

Nagasawa, M., Tanaka, H.: Propagation of Chaos for Diffusing Particles of Two Types with
Singular Mean Field Interaction
Probab. Theory Relat. Fields **71**, 69–83 (1986)

Nagasawa, M., Tanaka, H.: Diffusion with Interactions and Collisions Between Coloured
Particles and the Propagation of Chaos
Probab. Theory Relat. Fields **74** 161–198 (1987)

Nagasawa, M., Tanaka, H.: A Proof of the Propagation of Chaos for Diffusion Processes with
Drift Coefficients not of Average Form
Tokyo J. Math. **10**, 403–418 (1987)

Nagasawa, M.: Transformations of Diffusion and Schrödinger Processes
Probab. Th. Rel. Fields **82**, 109–136 (1989)

Nagasawa, M.: Stochastic Variational Principle of Schrödinger Processes
Seminar on Stochastic Processes (ed. Cinlar, Chung, Getoor), Birkhäuser, Basel 1989

Nagasawa, M.: Can the Schrödinger Equation be a Boltzmann Equation?
Northwestern Univ. 1989, in 'Diffusion Processes and Related Problems in Analysis'
(ed. Pinsky, M.), Birkhauser Boston Inc. (1990)

Nagasawa, M.: Schrödinger Equations and Diffusion Theory
Monographs in Mathematics 86, Birkhäuser, Basel 1993

Nelson, E.: Derivation of Schrödinger Equation from Newtonian Mechanics
Phys. Rev. **150**, 1076–1085 (1966)

Nelson, E.: Dynamical Theories of Brownian Motion
Princeton University Press, 1967

Neveu, J.: Mathematical Foundations of the Calculus of Probability
Holden-Day Inc., San Francisco 1965

Norris, J.R.: Construction of Diffusions with a Given Density
In 'Stochastic Calculus in Applications' (ed. Norris, J.R.), Longman 1988

Novikov, A.A.: On Moment Inequalities and Identities for Stochastic Integrals
Proc. 2nd Japan-USSR Symp. Prob. Th. Lect. Notes Math. **330**, 333–339
Springer-Verlag Berlin, 1973

Oelschläger, K.: A Martingale Approach to the Law of Large Numbers for Weakly Interacting
Stochastic Processes
Ann. Probab. **12**, 458–479 (1984)

Oelschläger, K.: A Law of Large Numbers for Moderately Interacting Diffusion Processes
Z. Wahrsch. verw. Gebiete **69**, 279–322 (1985)

Oelschläger, K.: Many-particle Systems and the Continuum Description of Their Dynamics
Habilitation, Univ. Heidelberg, 1989

Oshima, Y.: Some Properties of Markov Processes Associated with
Time Dependent Dirichlet Forms
Osaka J. Math. **29**, 103–127 (1992)

Parthasarathy, K.R.: An Introduction to Quantum Stochastic Calculus
Birkhäuser-Verlag, Basel 1992

Planck, M.: Zur Theorie des Gesetzes der Energieverteilung im Normalspectrum
Verh. d. D. Phys. Ges. **2**, 237–245 (1900)

Reed, M., Simon, B.: Methods of Modern Mathematical Physics I
Academic Press (1981)

Rogers, L. C. G., Williams, D.: Diffusions, Markov Processes and Martingales
J. Wiley & Sons Ltd, Chichester 1987; vol. 2

Schrödinger, E.: Was ist ein Naturgesetz?
Antrittsrede an der Universität Zürich, 9.12.1922
Sonderdruck aus 'Die Naturwissenschaften' **17**, 9–11, Springer-Verlag, Berlin 1929

Schrödinger, E.: Quantisierung als Eigenwertproblem, 1. Mitteilung
Annalen der Physik **79**, 336–376 (1926)

Schrödinger, E.: Quantisierung als Eigenwertproblem, 4. Mitteilung
Annalen der Physik **81**, 109–139 (1926)

Schrödinger, E.: Zum Heisenbergschen Unschärfeprinzip
Sitzungsberichte der Preussischen Akademie der Wissenschaften
physikalisch-mathematischen Klasse, 296–303 (1930)

Schrödinger, E.: Über die Umkehrung der Naturgesetze
Sitzungsberichte der Preussischen Akademie der Wissenschaften
physikalisch-mathematische Klasse, 144–153 (1931)

Schrödinger, E.: Sur la Théorie Relativiste de l'Électron
et l'Interprétation de la Mécanique Quantique
Ann. Inst. H. Poincaré **2**, 269–319 (1932)

Schwinger, J.(ed): Quantum Electrodynamics
Dover Publ. Inc., New York 1958

Sharpe, M.: General Theory of Markov Processes
Academic Press, Boston 1988

Shiga, T., Tanaka, H.: Central Limit Theorem for a System of Markovian Particles
with Mean Field Interaction
Z. Wahrsch. verw. Gebiete **69**, 439–459 (1985)

Simon, B.: Schrödinger Semigroups
Bull. Am. Math. Soc. **7**, 447–526 (1982)

Stroock, D. W., Varadhan, S. R. S.: Multidimensional Diffusion Processes
Springer Verlag, New York (1979)

Stummer, W.: The Novikov and Entropy Conditions of Diffusion Processes
with Singular Drift
Dissertation, Univ. Zürich, 1990

Sturm, T.: Schrödinger Operators with Arbitrary Nonnegative Potentials
Operator Theory: Advance and Applications **57**, 291–306 (1992)

Sznitman, A.S.: Non-linear Reflecting Diffusion Processes and Propagation of Chaos
and Fluctuations Associated
J. Funct. Anal. **56**, 311–336 (1984)

Sznitman, A.S.: Topics in Propagation of Chaos
École d'Été de Probabilités de Saint Flour, 1989

Tanaka, H.: Limit Theorems for Certain Diffusion Processes with Interaction
In 'Stochastic Analysis' (ed. Itô, K.), 469–488, Kinokuniya, Tokyo 1984

Tanaka, H.: Limit Distributions for One-dimensional Diffusion Processes
in Self-similar Random Environments
IMA hydrodynamic behavior and interacting particle systems (ed Papanicolaou, G.)
vol. 9, 189–210, Springer-Verlag, Berlin 1987

Tomonaga, S.: On a Relativistically Invariant Formulation of the Quantum Theory
of Wave Fields
Progress of Theoretical Physics 1, 27–39 (1946)

Varadhan, S.R.S.: Large Deviations and Applications
Soc. Ind. Appl. Math., Philadelphia 1984

Von Neumann, J.: Die Mathematischen Grundlagen der Quantenmechanik
Springer-Verlag, Berlin 1932

Wakolbinger, A: Los Procesos Estocásticos de Schrödinger
Ciencia 40, 199–208 (1989)

Wakolbinger, A: A Simplified Variational Characterization of Schrödinger Processes
J. Math. Phys. 27, 2943–2946 (1989)

Williams, D.: Diffusions, Markov Processes and Martingales
J. Wiley & Sons Ltd, Chichester 1979; vol. 1

Yasue, K.: Stochastic Calculus of Variations
J. Funct. Anal. 41, 327–340 (1981)

Yasue, K.: The Least Action Principle in Quantum Theory
Soryusiron Kenkyu (Japanese), 1986

Yoshida, K.: Functional Analysis
Springer-Verlag, Berlin 1965

Yukawa, H.: On the Interaction of Elementary Particles I
Proc. Phys.-Math. Soc. Japan 17, 48–57 (1935)

Zhao, Z.: Schrödinger Conditional Brownian Motion and Stochastic Calculus of Variations
Stochastics 18, 1–15 (1986)

Zambrini, J.C.: Stochastic Mechanics according to Schrödinger
Phys. Rev. 33, 1532–1548 (1986)

Zheng, W. A.: Tightness Results for Laws of Diffusion Processes.
Application to stochastic mechanics
Ann. Inst. Henri Poincaré B21, 103–124 (1985)

Index

PA - Probability and its Applications

Edited by
Th. M. Liggett / Ch. Newman / L. Pitt

Probability and its Applications publishes research-level monographs and advanced graduate texts dealing with all aspects of probability theory and stochastic processes, as well as their connections with and applications to other areas such as mathematical statistics and statistical physics.

R. Carmona/J. LaCroix
Spectral Theory of Random Schrödinger Operators
1990. ISBN 3-7643-3486-X

R.K. Getoor
Excessive Measures
1990. ISBN 3-7643-3492-4

K.L. Chung/R.J. Williams
Introduction to Stochastic Integration
1990. ISBN 3-7643-3386-3

G.F. Lawler
Intersections of Random Walks
1991. ISBN 3-7643-3557-2

H. Linhart/W. Zucchini
Statistik Eins
1991. ISBN 3-7643-2586-0

R.M. Blumenthal
Excursions of Markov Processes
1992. ISBN 3-7643-3575-0

S. Kwapien/W. Woyczynski
Random Series and Stochastic Integrals: Single and Multiple
1992. ISBN 3-7643-3572-6

N. Madras/G. Slade
The Self-Avoiding Walk
1992. ISBN 3-7643-3589-0

PP - Progress in Probability

Edited by
Th. M. Liggett / Ch. Newman / L. Pitt

Progress in Probability is designed for the publication of workshops, seminars and conference proceedings on all aspects of probability theory and stochastic processes, as well as their connections with and applications to other areas such as mathematical statistics and statistical physics.

BIRKHÄUSER *MATHEMATICS*

M. Nagasawa, Universität Zürich, Switzerland

SCHRÖDINGER EQUATIONS AND DIFFUSION THEORY

MMA 86 – Monographs in Mathematics

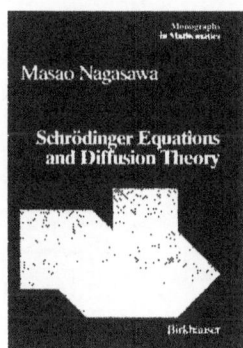

1993. 332 pages. Hardcover
ISBN 3-7643-2875-4

Schrödinger Equations and Diffusion Theory addresses the
question "What is the Schrödinger equation?" in terms of
diffusion processes, and shows that the Schrödinger equation
and diffusion equations in duality are equivalent. In turn,
Schrödinger's conjecture of 1931 is solved. The theory of
diffusion processes for the Schrödinger equation tells us that
we must go further into the theory of systems of (infinitely)
many interacting quantum (diffusion) particles.
The method of relative entropy and the theory of transforma-
tions enable us to construct severely singular diffusion processes which appear to be equivalent to Schrödinger equations.
The theory of large deviations and the propagation of chaos of interacting diffusion particles
reveal the statistical mechanical nature of the Schrödinger equation, namely, quantum
mechanics.
The text is practically self-contained and requires only an elementary knowledge of probability theory at the graduate level.

*"There is a long standing controversy on the meaning of Schrödinger equations. Nagasawa
devotes a whole appendix to approaches by Fényes, Nelson, Zambrini, Bohm, and others. He
manages to discuss them all in the framework of his approach and he clearly works out their
main defects. ... The precise and plentiful explanations aid the readers who do not have an
infinite amount of time; other readers have the chance to enjoy some really difficult proofs
employing stochastic analysis."*

SIAM, Vol. 36, 6/94

**Please order through
your bookseller or:**
Birkhäuser Verlag AG
P.O. Box 133
CH-4010 Basel / Switzerland
FAX: ++41 / 61 / 271 76 66
e-mail: 100010.2310@compuserve.com

**For orders originating in
the USA or Canada:**
Birkhäuser
333 Meadowlands Parkway
Secaucus, NJ 07094-2491
USA

Birkhäuser

Birkhäuser Verlag AG
Basel · Boston · Berlin

**Please contact us for being regularly informed
about our latest publications in this field.**